A NATION
FORSAKEN

A NATION FORSAKEN

EMP: THE ESCALATING *THREAT* OF AN AMERICAN CATASTROPHE

F. MICHAEL *MALOOF*

 WND BOOKS

A NATION FORSAKEN

WND Books
Washington, D.C.

Book designed by Mark Karis

WND Books are distributed to the trade by:
Midpoint Trade Books
27 West 20th Street, Suite 1102
New York, New York 10011

WND Books are available at special discounts for bulk purchases. WND Books, Inc., also publishes books in electronic formats. For more information call (541) 474-1776 or visit www.wndbooks.com.

First Edition

Hardcover ISBN: 978-1-936488-56-8
eBook ISBN: 978-1-936488-92-6

Library of Congress information available

Printed in the United States of America
10 9 8 7 6 5 4 3 2 1

TABLE OF CONTENTS

FOREWORD

YEARS BEFORE U.S. INTELLIGENCE AGENCIES began focusing on rogue nations obtaining nuclear arsenals, Michael Maloof foresaw the potential threat. His vision was clear. His warnings loud. His advice . . . too often dismissed.

In the ensuing years, a variety of nations hostile to the United States have, as Michael predicted, been racing to mount a nuclear device on a long-range missile—directly threatening American soil. The doubters/ skeptics who once dismissed Michael's warnings as inconsequential and uninformed have now ceded to Michael's ominous warnings—that America is surrounded by political despots and religious fanatics who have gained the means to disrupt our society in ways never before imagined.

Michael earned a reputation—not always appreciated—for speaking truth to power when he served as senior policy analyst at the Department of Defense. He cares little for the entrenched bureaucracies beholden to political expediency, unseen corporate masters, or self-interest. Michael has always done what is necessary to protect American lives—he is a real-life Jack Bauer in a time when twenty-four hours can dramatically upend all we hold sacred.

Long before the terms "terrorist" and "jihad" entered the American lexicon, before American policymakers and leaders came to recognize the threat Afghanistan poses to the United States, Michael had already laid the groundwork in Central Asia to train border guards in an effort to help build barriers against nuclear proliferation and what then was emerging as an increasing threat—terrorism.

Michael's aggressive methods for halting the flow of Western nuclear and missile technologies to our enemies was often derided at the highest levels of government. But to those of us working in Congress, including members of the House Armed Services Committee, who listened in rapt horror at Michael's revelations, he was, and is, a hero.

Michael is going to convince you as well, I believe, that there are state sponsors of terror who now have the capability to bring the United States to its knees. And not just the capability, but also the desire. You will learn how a nuclear weapon, exploded over our country, can deliver an electromagnetic pulse that literally fries every electronic device within sight, triggering the collapse of all communications, electricity, transportation, financial services, food, and water supplies . . . leaving 313 million Americans in a state of anarchy. And you will see as well that America's leaders have known about this threat for decades.

The Congressional EMP Commission, on which I served, did an extensive study of this electromagnetic pulse—this EMP, as it is known. We discovered to our own revulsion that critical systems in this country are distressingly unprotected. We calculated that based on current realities, in the first year after a full-scale EMP event, we could expect about two-thirds of the national population—200 million Americans—to perish from starvation and disease, as well as anarchy in the streets.

That's how unprepared we are.

Mother Nature also poses her own violent EMP threat. The sun's well-known solar flares can trigger intense geomagnetic storms on Earth if they strike the right—or rather, wrong—way. In fact, every century or so we've experienced a geomagnetic super storm strong enough to fully collapse the electrical grids and critical infrastructures everywhere on Earth, requiring years to recover—if recovery is possible at all.

It has happened before. Michael will show you why it's just a matter of time before it happens again.

Another kind of EMP threat is posed by radio-frequency (RF) weapons that can be made by terrorists or criminals using parts purchased at Radio Shack. RF weapons do not have the power or range of a nuclear

EMP attack. But an RF weapon used intelligently could cause a localized disaster, perhaps blacking out a major metropolitan area, or worse. RF weapons make it possible for the first time in history for a single terrorist, criminal, or madman to topple the pillars of civilization in an entire city.

Michael performs a great public service with this book. He has built on the findings of the EMP Commission and other important studies, and for the first time rendered them accessible to the general public in an easily understandable way.

Despite the staggering enormity of these EMP and RF threats, there are solutions and safeguards available to us. Michael will show that by increasing the average electric bill just twenty cents annually, we can protect the national electric grid. He will also show you why it is not being done.

In fact, the government has joined with private industry to pretend this threat "isn't so bad" and that "we needn't worry."

Nothing could be further from the truth.

Common sense tells us that the electric power industry may not be the most trustworthy steward of public safety—if it could result in a revenue shortfall for them. Trusting them would be analogous to trusting the zeppelin industry in the 1920s when they claimed that zeppelin travel using flammable hydrogen was perfectly safe—just prior to inflating the Hindenburg.

I've spent much of my professional career trying to protect the United States from an EMP catastrophe—first at the CIA, on the House Armed Services Committee, on the Congressional EMP Commission, on the Congressional Strategic Posture Commission, and now as executive director of the Task Force on National and Homeland Security. I give my highest recommendation to this book, and salute the ever-vigilant Michael Maloof.

—*Dr. Peter Vincent Pry, Executive Director, Task Force on National and Homeland Security*

1

A DIRECT ATTACK ON OUR NATION'S CAPITAL

A BAD COMMUTE MADE WORSE

It's a sunny Tuesday morning, with the usual rush-hour traffic in the nation's capital, home to the federal government. Ten bridges link neighboring Maryland and Virginia with Washington, D.C., carrying almost half a million commuters into the city each day. Radio announcers groan about traffic backing up.

A car has stalled on the 14th Street Bridge—not far from the Pentagon. A few minutes later, a second car stalls on the same bridge. Then two more cars stop in the middle of traffic on the Roosevelt Bridge—which also brings a high volume of traffic in from Northern Virginia. Then more news breaks. Cars are stalling on the Key Bridge, as well, and on the 11th Street Bridge, and the South Capitol Street Bridge. Cars have also stopped in key intersections along Constitution Avenue, south of the White House, choking this main artery into the city. With traffic at a standstill, explosions are heard rocking the L'Enfant Plaza, Metro Center, and Union Station subway stations. Nobody has any idea what's going on. And no one can move.

Emergency vehicles are unable to respond due to traffic jams spanning the length and breadth of Washington.

Tens of thousands of people—mostly federal employees—take to their cell phones to find out what has happened. But cell coverage, where it works at all, becomes quickly overloaded. Longtime Washingtonians recall having the same problem in the wake of 9/11. This is going to be a long day.

Then, in a delivery box truck left idling at the curb just a few blocks from McPherson Square and Lafayette Park, an electromagnetic bomb is triggered, emitting a high-energy pulse that in a blink disables the downtown electrical grid within a four-block area.

Half a dozen federal agencies are incapacitated. Especially hard-hit are the Old and New Executive Office buildings and the Treasury Department, flanking the White House to the west, north, and east, respectively.

Hundreds of cars and vans within the area jerk to sudden stops. Passengers on Metro buses are thrown violently out of their seats as the buses bounce hard and shut down. Subway trains below ground are immobilized as their brakes engage mechanically when the power to the third rail fails.

A D.C. police helicopter, flying low over the traffic tie-ups near the White House, takes a portion of the electromagnetic pulse directly into its engine's power circuitry, causing the turbine to fail and sending the chopper careening into the crowded streets below.

Two blocks from the White House, on a tour bus filled with aging veterans in town to visit the World War II and Vietnam memorials, two men clutch their chests as their pacemakers explode, and another twenty war heroes suffer brain injuries when their hearing aids take the full force of the electromagnetic shockwave.

The pilot and copilot of a United Airlines 757, on its northbound departure from Reagan National Airport, watch helplessly as their flight management screens fail and their navigation and systems displays go black and then try to reboot. Other electrically operated systems on the plane seem to surge or die. Only the pilot's reflexive move to switch to battery power prevents the plane from plowing into the Washington Monument and National Mall.

Radio and television stations attempt to interrupt local programming to bring breaking news of a "potential terrorist attack." But no luck. The heart of the city is electronically dark in broad daylight. The networks' key anchor booths built atop the Hay-Adams Hotel opposite the White House are rendered useless. The microwave antennas no longer beam their signals to the network control centers.

Reporters in the middle of their morning stand-ups on the White House north lawn lose their video and audio feeds in a blink. Sound technicians with headsets are slammed with eardrum-crushing screeches as the electromagnetic pulse courses through the headphone wires. News of the attack—if that's what it was—only trickles out of nearby bureaus not affected by the blast.

Pedestrians hooked up to their iPods, iPhones, BlackBerrys, and Androids find themselves holding useless metal rectangles.

Across the Potomac River, in the Pentagon, desk officials notice a sudden drop-off in the data stream normally coming from the White House. They also notice this happened at about the same moment the explosions were heard. Security alerts ring on all five sides of the Pentagon.

In the White House itself, the president and senior staff are moved boldly to the President's Emergency Operations Center (PEOC) beneath the East Wing. There, a team of military and civilian briefers are furiously trying to figure out what the hell is happening.

By noon, Washington is a chaotic mess, as powerful people used to getting their way are now mere commoners, and afraid. Across the city, people are in trouble. Thousands of injured residents, workers, and tourists are being physically barred from hospitals and clinics, already overflowing, and without power they cannot properly function anyway. The federal government is operating on a bare-bones emergency planning basis. Local police and a dozen or more federal law enforcement agencies and city service operations are overwhelmed.

Yes, this is going to be a long day.

HOW COULD THEY PULL OFF SOMETHING LIKE THIS?

The short answer is, "Quite easily."

The cars on the bridge were fired upon by special weapons—small, rifle-sized arms that shoot not bullets but radio frequencies, weapons that can be built for about $400 with easy-to-obtain parts. Think of one of those Super Soaker water guns. The operating principle is simple. Water in a container. Air pressure pumped up in a chamber. Combine water and the high-pressure air in a tube, and out comes a powerful squirt. With a radio-frequency (RF) device, it's similar. Begin with a magnetic field in a tube that's wrapped with a copper coil and surrounded by capacitors. Charge up the capacitors. Then add in a good energy source, such as an ignited C-4 explosive, and—*whooof!*—you're firing a stream of highly charged atomic particles toward an electronics-based target (your car's electronics, your iPhone, your pacemaker, your national security communications . . . you get the point). Unlike the water gun's targets, things damaged by an RF weapon cannot be dried off. They are killed off, unusable until replaced.

In this Washington scenario, the terrorists' strategy was to take advantage of the potential choke points created by the predictable traffic influx into the city. Halting traffic on the bridges and inside the city prevented emergency vehicles and first responders from reaching areas where bombs, carried onto the subway system in backpacks, had exploded, sending a message not only of hate but, perhaps more frightening, of the fact that they could do this again and again until the "Great Satan" is brought to its knees. That's the message sent via the electromagnetic bomb, just blocks from the White House: *We may be small; we may be unsophisticated by your standards, but we can hurt you whenever we want to.*

THE EASIEST WAY TO BRING DOWN AN AIRLINER

At Reagan National Airport, a nondescript minivan parks in a public lot just north of the airport fence. A couple emerges with a picnic lunch and sits in a grassy spot beneath the low-flying planes. At the same time, a pickup truck towing a small boat arrives at the launch that's two hundred yards from the main runway. Two men then launch a Boston Whaler.

Both of these teams of terrorists have brought RF weapons to the airport in plain sight—one in a large picnic basket, the other in a beer cooler.

A third terrorist team strolls onto Roosevelt Island, a National Park Service-administered plot of paradise situated immediately beneath the Reagan National flight paths. This team sets up the third radio-frequency (RF) weapon here, under the canopy of trees and easily aimed at aircraft navigating their way along the Potomac River's course.

Each team monitors the publicly available radio transmissions between the control tower and departing airliners.

As one heavy jet departs and begins its climb to the north, another jet begins its takeoff roll. On Roosevelt Island, the terrorists take aim at the first plane—now climbing through 1,500 feet directly above them. At the park near the runway, the couple with the picnic basket and the men in the boat are poised to trigger their electromagnetic bombs as the second plane lifts off the runway and soars just 200 feet over their heads. Simultaneously, all the teams fire their weapons—sending high-energy electrons streaming into the planes, jamming everything from the navigation systems to the electronically directed engine controls.

At the cost of a few thousand dollars in materials and know-how, this homegrown terror cell kills more than a thousand people—several hundred passengers on the planes, the rest in the buildings that take the full impact of the crashing planes.

IS THIS TRULY POSSIBLE? IT HAS ALREADY BEEN CONTEMPLATED.

Unlike the RF device unleashed in downtown Washington, an RF weapon aimed at commercial air traffic at an aviation choke point would be catastrophic. Commercial airliners all enter a crucial phase of flight during which even modest deviations from speed and altitude can have unacceptable—usually fatal—consequences. The big jets—Boeings and Airbuses—are fuel-laden and often at maximum weights as they climb over densely populated areas at vulnerable speeds, clawing their way to safer and faster altitudes.

This is a crucial time. That is why flight attendants instruct us to turn off our electronic devices prior to takeoffs and landings. Low levels of electronic interference from laptops, cell phones, and e-book readers can have some small influence up in the cockpit. So imagine the impact of a massive pulse of electromagnetic energy on a plane. Visualize that focused burst of energy being triggered by a terrorist who knows full well just how tenuous is the link between aircraft control and human disaster.

A plane that is hit and disabled by an RF wave generated on Roosevelt Island would most likely crash in Georgetown or Rosslyn. The swath of destruction caused by a 757 with nearly 80,000 pounds of fuel on board is incalculable. The terrorists are counting on that.

WHAT IS GOING ON HERE?

The force that brought downtown Washington to a standstill, cut communications networks, and damaged and destroyed a wide range of electronic devices, from automobile engine computer chips to implanted pacemakers, is known as an *electromagnetic pulse*, or *EMP*. An EMP is a high-intensity burst of electromagnetic energy caused by rapid acceleration of charged particles. In the form we have seen so far—the radio-frequency e-bomb, let's call it—an EMP can be generated on the cheap by almost anyone with a bit of knowledge and access to some relatively simple components.

The bigger brother of EMP is the massive pulse generated by a nuclear explosion many miles above Earth. The granddaddy of EMP production is the sun, normally capable of a bad burn at the beach, but sometimes capable of frying satellites, shutting down GPS systems, and paralyzing entire continent-spanning electric power grids.

None of this is science fiction; it is science fact. The nation's top scientists and analysts, our military leaders, and our policymakers, from Capitol Hill to the Oval Office, are all quite familiar with EMP and its potential for disruption and destruction. The power of EMP is also well known by our allies and our enemies—from nation-states to terrorist cells.

Open-source information has documented how an EMP can be used against aircraft in what's known as an *intentional electromagnetic interference pulse*. In a 2005 technical paper titled "Potential IEMI Threats against Civilian Air Traffic," D. J. Serafin outlined such a scenario: "An airport area could be a selected target for EM [electromagnetic] terrorism due to the high concentration of electronics equipment likely to be perturbed by EM threats, so producing broad chaos."[1]

Serafin then provided two scenarios for introducing a small RF weapon: one concealed inside a suitcase and placed near terminal computer networks; the other a truck-mounted RF weapon, which could be located near an airport, with direct view of the runways and with a range extending nearly a half mile.

In his scenario concerning the introduction of RF weapons into the airport, Serafin provided detailed descriptions of the microwave bandwidth, distance, and megahertz ranges for the most effect—something a technically competent terrorist would easily understand and duplicate.

Targets for the RF weapon would include such aircraft equipment as onboard navigation and global positioning systems, distance measuring equipment, traffic collision avoidance systems, and the air traffic control system. Because of the antenna situated on top of the aircraft's fuselage, these systems would be vulnerable.

RF PENETRATES WALLS, JUST LIKE A RADIO

Electromagnetic energy is so powerful it can penetrate walls and other barriers in a facility, such as windows. However, the more walls that have to be penetrated, the harder it is for RF weapons to be effective on targeted electronics. In addition, distance matters. The farther the source of the RF signal from a target, the less damage to the electronics. And if the targeted electronics are enclosed within barriers—especially walls and a conductive shield—it becomes even more difficult for the RF to interfere with them.

Unfortunately, even a facility that has been fully "shielded" can remain quite vulnerable. An attacker can attach a conductive RF weapon to a power line that's miles away from the target, knowing that power line leads straight into the targeted facility. When an RF pulse is directed into the line, the energy zips along the power line, destroying anything electronic along the way, instantly.

EMP AS A SUITCASE BOMB

An EMP shock wave can be produced by a device small enough to fit in a briefcase. The technology has matured to the point where practical e-bombs are technically feasible, with new applications in both strategic and tactical warfare. What worries terrorism experts most is the *flux compression generator*, or "flux gun" for short.[2] This is a rather simple weapon under development in the United States and other countries, including rogue states, as we will see.

The flux gun consists of an explosives-packed tube placed inside a slightly larger copper coil. The instant before the explosive is detonated, the coil is energized by a bank of capacitors, creating a magnetic field. The explosive charge detonates from the rear forward. As the tube flares outward, it touches the edge of the coil, thereby creating a moving short circuit. This produces a ramping current pulse that makes a lightning bolt look as lowly as a flashbulb. We'll talk more about this in coming pages.

Although much of the U.S.-based work on the flux gun is classified, it is believed that current development efforts are based on using high-temperature superconductors to create intense magnetic fields. This flux gun is powerful enough to cripple electronic wiring and circuitry over a geographic area of several square miles, posing a real threat to the nation's critical infrastructure.

CAN BE MADE FOR $400 FROM SCHEMATICS FOUND ON THE INTERNET

It took 130,000 of our nation's finest minds a full six years and $26 billion in today's dollars to perfect a nuclear bomb. It takes passing grades in high school science and $400 to create an RF weapon capable of doing just as much damage, if deployed strategically. That is the cold, infuriating truth of this new threat. A quick Internet search will yield a number of "how-to" sites with complete schematics for constructing a simple but effective EMP weapon using components found at any electronics store.

One weapons specialist set out to prove just how easy it is. The company, Schriner Engineering, published its findings in the May 3, 2001, *New Scientist.*

Schriner was testing to see if cheap, homemade radio-frequency weapons could be built by people with little technical know-how. These tests were conducted on behalf of the Department of Defense. All legitimate.

The tests were successful.

The RF weapons were made from components readily available from electronic stores and catalogs. The weapons generated an extremely short but powerful pulse of electromagnetic energy. Mike Powell of Schriner Engineering said that such RF weapons would be capable of bringing down an aircraft, something that would be of particular concern in Washington due to the intricate and restricted flying route large commercial aircraft must take on their approaches to or departures from Reagan National Airport.

"The message here is that any number of groups in the U.S. or other countries can do just this, relatively easily and at relatively low cost," Powell said in a May 3, 2001, *New Scientist* article.

As David Schriner of Schriner Engineering concluded, "Our whole nation is vulnerable . . . We dance along with all this high technology, and we're very dependent on it. But if it breaks, where will we be?"[3]

Interestingly, Schriner brought an RF device to the U.S. Capitol for "show-and-tell" when he was testifying there. The sergeant at arms learned what the device was capable of—namely, frying the electronics of computers in all the Capitol office buildings—and the device was not allowed in. In his testimony, Schriner told how he made the RF device in his own garage. His point was to show that the low-end technology needed to piece together an RF weapon was available without effort, at a very reasonable cost.[4]

A year earlier, General Robert Schweitzer had told members of Congress of a similar challenge he had undertaken. He instructed a group of young scientists from a national laboratory to build a working RF weapon. The aspiring scientists went down to their local Radio Shack, bought the components needed to make the weapon, and mounted it on top of a minivan. Schweitzer summed up what could have happened next:

> So, you've got a situation . . . where you could put components from Radio Shack inside of a van no bigger than a UPS truck with an antenna. And, that's really what an RF weapon often looks like, a radar or antenna showing, and drive it around the Dirksen Building, make a series of passes over the Pentagon or the White House, or the FAA facility out at Langley . . . You make a number of passes around the building and emit these pulses. They go through concrete walls. Barriers are no resistance to them. And, they will either burn out or upset all of the computers or the electronic gear in the building.[5]

With a radar loaded in the back of a van or pickup, an RF weapon can be directed at whatever target is intended. Because the radar is directional, it won't have any effect on the vehicle carrying the radar, as long as it is pointed away from its electronics.

RF weapons have been used to defeat security systems, disable police communications, and disrupt bank computers. More advanced RF weapons can jam satellites, trigger aircraft crashes, create pipeline explosions and large gas spills, and cause lifesaving medical equipment to malfunction. They can also be used to make public water systems malfunction, potentially resulting in flooding. Engineering is on our side. Others are not. And for them there are a number of websites that peddle RF devices out in the open, almost flaunting the risks involved.

The website http://blockyourid.com/~gbpprorg/mil/emp/index.html, for example, offers a step-by-step approach. This is not an endorsement of the website—far from it. It is simply illustrative of how easily this kind of bomb-making information can be obtained.

The RF device is not capable of taking out every computer in town. But the website does provide a technical overview of the steps needed to conduct EMP warfare, including a list of the off-the-shelf components that go into making an RF generator. The most critical component, says the writer, is a high-voltage pulse capacitor. For "really significant results," it will need "a pulse capacitor output in the hundreds of Joules." The website also offers suggestions on boosting the electromagnetic pulse. The price of adding capacity to the EMP generator does begin to add up, but it's still relatively inexpensive.

For any garden-variety terrorist who doesn't want to build an RF weapon, there are a number of websites to turn to for the finished product. One example is Information Unlimited (www.amazing1.com), which offers an entire catalog of EMP devices. This is the sales blurb on their "EMP/HERF/SHOCK Pulse Generator":

> Shock wave generators are capable of producing focused acoustic or electromagnetic energy that can break up objects such as kidney stones and other similar materials. EMP generators can produce pulses of electromagnetic energy that can destroy the sensitive electronics in computers and microprocessors. Destabilized LCR circuits can produce multi megawatt pulses by using an explosive wire disruptive switch. These high power pulses can be coupled "with difficulty" into antennas, conic sections, horns etc., for very directional effects. Research is currently being undertaken to disable vehicles thus avoiding dangerous high speed chases. The trick is to generate a high enough power pulse to fry the electronic control processor modules without creating collateral damage to unintended targets. This could be a lot simpler if the vehicle was covered in plastic or fiberglass rather than metal. The shielding of the metal body offers a challenge to the researcher to develop a practical system. A system could be built that could do this but would be costly, large and produce collateral damage to friendly targets. Devices described

are intended for experienced researchers and qualified personnel who are aware of all hazards and liabilities in use of this equipment.[6]

There can be no misreading this sales material. The vendor is implying guidance on amplifying the systems to provide greater damage to an intended target. The catalog then goes into the specific generators, some of which come with the warning: *"This is an advanced and highly dangerous project!"*

Oh, really? How about two exclamation points, then?

Another of Information Unlimited's featured offerings is "Complete EMP System Crated with All Setup Instructions." The company makes it clear that they'll only sell to qualified research outfits. But how difficult is it for any number of buyers to mask their true identity and intent? Not hard at all.

IT'S A MARKETPLACE FOR MAYHEM

In breezing through the company's online catalog, you see the EMP150 selling for $8,000, weighing about 200 pounds, capable of instantly pulsing 300 megawatts at a target. For more ambitious mayhem, the EMP400 is good for a 1.8 gigawatt output. And for serious terrorists, $32,000 gets you the "Electromagnetic EMP Blaster Gun, Gen II." This charmer is capable of shutting down a bank of computers or blowing up a gas station—from a distance of 50 feet.[7]

With some technical savvy, a person can take these basic elements and boost them to even greater levels of lethality. All the required products—lasers, capacitors, rectifiers, diodes, etc.—can be ordered off the website just like ordering clothing from a catalog. A single disclaimer puts the burden of responsibility on the buyer, not the seller:

Legal Status of These Products: It is the responsibility of the buyer, not the seller, to ascertain and obey all applicable local, state, federal, and international laws in the regard to the possession, use and sale of any item purchased from Information Unlimited, Inc. Absolutely no sales to minors. By placing an order, the buyer represents that the products ordered will be used only in a lawful manner and that he/she is of legal age. Information Unlimited, Inc. will not be held liable for the misuse of any product purchased from us or any of our direct distributors or dealers. Also, whereas no product is 100% effective against attack, Information United Inc. assumes no responsibility if a personal safety product purchased from us is not effective in preventing bodily injury or death. Some of the items we carry are "DANGEROUS" and may be "ILLEGAL" in your state or country. Please be sure to check local laws "BEFORE" ordering.[8]

For some of the items in the catalog, Information Unlimited requires the signing of a "Hazardous Equipment Affidavit." The form says that "hazardous devices are only made available to qualified and established companies in these related fields."[9] That's it. No explanation of what constitutes a "qualified" company. And when contacted for details on how qualifications are determined, the company did not respond.

EASIER THAN BUYING A GUN

Mail-order catalogs that sell ammunition require the buyer to, at least, submit a photocopy of a driver's license or ID card, complete with photograph, to establish a paper trail and bona fides. No such information is needed for acquiring RF weaponry.

In the event of an EMP attack, this basic buyer information would be genuinely useful to FBI investigators. It might produce a lead. However Information Unlimited is headquarters in the "Live Free or Die" state of New Hampshire, and doesn't have such a requirement. That said, it would be difficult under the best of circumstances to keep RF weaponry out of the hands of bad actors globally. None of the components to make these weapons is subject to export controls. In fact, all of the components have been *de*controlled, making the technology easily obtainable on the open market to anyone who wants to make an RF weapon.

RF WEAPONS COME IN ALL SIZES AND PACKAGES

RF weapons are produced to many different specifications—depending on the intended end use. They can be handheld devices, such as a briefcase or even a package or small soda can, or large weapons that can be loaded into a vehicle—a van, utility truck, delivery or commercial vehicle, even a pickup—and placed outside a facility.

In an urban environment, a delivery truck could be parked outside a target building and go unnoticed for an extended period while doing its destructive work. On any high-traffic street in a major metropolitan area such as New York City, Washington, D.C., or Boston, for example, it isn't unusual to see a large number of delivery vehicles—many unmarked—outside of busy commercial centers. Once the weapon is fired—and even that can be done from a remote location—all electronics in the targeted buildings will be destroyed. And the vehicle can then be booby-trapped to detonate at a prescribed time—to destroy evidence.

An attacker may choose to use these smaller RF weapons over conventional weapons for a variety of reasons. An RF attack can be covert, with the attacker achieving the desired goal without leaving a trace. The only evidence is the sudden malfunctioning or total breakdown of electronic equipment. Particularly vulnerable are telephone and utility companies where the relay lines are exposed, with only minimal protective fencing.

When an attacker cannot approach a stand-off target—for example, a well-guarded nuclear facility—attacks can be launched remotely by simply aiming at the target from a distance. The aim does not have to be precise—a strike anywhere close to exposed electronics will accomplish the objective.

Another advantage of RF weapons is the ammo—they don't use any. They require only a power source. The heftier the power source used to generate the electromagnetic energy, the greater the disruption and destruction.

RF energy can penetrate walls and move around obstacles. Once an RF weapon is used, the damage or destruction to vital electronics will be instantaneous, since RF energy travels at the speed of light. Exposure can also be timed to a specific moment. A TV news station can be taken down just before broadcasting a story about surviving an EMP attack, for instance.

BUCK ROGERS, YOUR FLUX GUN HAS ARRIVED

We saw how destructive an electromagnetic pulse could be in the Washington scenarios. Let's look closer at the radio-frequency (RF) threat.

Radio-frequency weapons are classified by the type of beams they create: *wideband* or *narrowband*. Wideband RF weapons work in low-frequency ranges from 10 megahertz to 1 gigahertz, creating short pulses. Narrowband weapons operate at a single frequency from 1 gigahertz to 35 gigahertz, creating longer pulses.

If there's any weapon that comes close to the ray gun Buck Rogers made famous, it's the flux compression generator mentioned earlier. This weapon was invented by A. Sakharov in Russia and C. M. Fowler of Los Alamos National Laboratories. It's the most powerful "energy gun" ever built, capable of firing an electromagnetic pulse of tens of megajoules— which is roughly equal to a ten-ton truck slamming into you at 100 miles per hour. It does this faster than you can blink. And the weapon practically fits in a holster. It's a genuine "weapon of mass destruction." Think, a billion watts in a billionth of a second—that's the profile of this flux gun. And the weapon's range is 600 to 3,000 feet—so a terrorist could aim at a target from an easy distance, keeping out of sight, avoiding detection.

The best protection against the flux gun is something known as a *Faraday cage*, designed in 1836 by Michael Faraday to protect against lightning strikes. But the Faraday cage is not 100 percent reliable against narrowband RF (low-frequency, short pulses). The cage's solid metal exterior will repulse an RF attack. But a directed pulse from an RF weapon is so naturally overwhelming that any wiring leading into the cage acts as an antenna, carrying the pulse directly into the electronics the structure was meant to protect. Because the RF frequency is very high—which means the energy waves are tightly packed together and capable of penetrating very small openings—even a tiny crack or hair-thickness seam in the protecting shield can be breached and cause serious damage.

In a nonhostile environment, the Faraday cage is adequate protection. But in a hostile situation, it has its limitations. Most electronic equipment is still wired to the outside world, and this provides an entry point for an electromagnetic current, which can then run up the lines into the protected equipment, destroying everything in its electronic path.

Even if a device is perfectly shielded, it can degrade over time through vibration, corrosion, and breakdown of gaskets. With only the slightest wear and tear, an RF signal could soon be penetrating the shields and damaging the electronics. A "hardened" system can begin to break down in just six months. This means that hardening is not a onetime effort, but must become a steady, ongoing effort.

OUR MILITARY'S FRONT *AND* BACK DOORS ARE VULNERABLE

As technology has marched on in the military arena, with electronics becoming more densely packed, more energy efficient, and operable at higher speeds, these electronics become even more susceptible to RF microwave radiation.

RF radiation has two means of entering a target: front-door and back-door coupling. *Front-door coupling* occurs when RF radiation enters a system through a sensor or antenna designed to receive this type of radiation. The antenna can actually amplify the signal and allows for the energy from the RF radiation to pass directly to internal components. *Back-door coupling* allows RF radiation to sweep in through cracks, seams, cables, solar cells, optical sensors, and the like. Ultra-wideband radiation is especially effective for back-door coupling because of the wide range of wavelengths that allow for penetration from multiple locations on the target.

Any asset of our military that can be enclosed in a perfectly seamless case made of conducting metal with no sensors or connections to the outside—in other words, a Faraday cage—can be protected. All the electronics inside can be totally shielded from external signals. The reality, however, is that it is impossible to wrap automobiles and aircraft in such a cage. Aircraft and precision weapons have sensors, flight control surfaces, and other openings to allow RF radiation to enter and wreak havoc.

In an actual wartime scenario, our military would know how vulnerable they are. They would know that our very sophisticated precision-guided standoff weapons could be compromised by RF weapons. The missiles may fire; they may follow the programmed trajectory; they may hit their intended targets. Or they may not. And in not knowing for certain, our military commanders would be forced into an aggressive posture in protecting the homeland. Our armed forces are accustomed to being the defender, not the aggressor. But to win in this new form of warfare, we may have to launch surprise attacks against an adversary's communications and integrated air defense systems if we hope to gain an initial and hopefully lasting advantage.

2

THE EASTERN UNITED STATES— SOMEDAY SOON

ABOUT 120 MILES EAST-SOUTHEAST of the mouth of New York Harbor, the relatively shallow seabed of the continental shelf borders the deep valleys and trenches of the Atlantic. There, at a place oceanographers have named the Babylon Canyon, a lone modified *Sabalo*-class variant of the German-made Type 209 attack submarine rises from the depths, surfaces, and stabilizes its position. In less than a minute, a slender missile erupts out of a specially designed, deck-mounted launching tube, and arcs gracefully toward the northwest into a late fall dusk.

Rising higher and higher, the missile curves across the U.S. coastline, passing far above Red Bank, New Jersey. Still in the low atmosphere, over Reading, Pennsylvania, the missile's first stage burns out, and the second-stage engine kicks in, urging the rocket to an altitude of about 100 miles, at a speed approaching Mach 5, or 3,750 miles per hour. As the missile reaches the apogee of its ballistic flight slightly north of Pittsburgh, a computer program in the heart of the rocket, having oriented itself using American GPS signals, commands the one-megaton nuclear warhead riding up front to detonate. The cities below are covered in a thin overcast, with no forecasts of storms, but people on the ground will recall seeing what they'll describe as white-green sheet lightning illuminating the clouds from above. There will be no reports of thunder. Witnesses will also recall that at the moment the clouds lit up, the city of Pittsburgh went dark.

Four hundred miles due east, the evening commuters heading out of Manhattan back to New Jersey, Connecticut, and any of the New York boroughs suddenly find themselves behind the wheels of tens of thousands of stuttering, jerky, and powerless cars, vans, and SUVs, staring at a lightless metropolis. As far as the eye can see, streetlights, traffic signals, apartment and office buildings, and billboards are all dark. Not even a bit of light can be detected. The East Coast has gone dark.

THE STONE AGE IS JUST AROUND THE CORNER

In the faint glow of a cold, October evening, New Yorkers—like their fellow citizens throughout the Northeast and Midwest—face a crisis of never-imagined proportions. This clearly isn't the Northeast blackout of November 1965 which affected parts of Ontario, Canada, Connecticut, Massachusetts, New Hampshire, Rhode Island, Vermont, New York, and New Jersey, leaving some thirty million people without electricity for some twelve hours.

Instead, this is something far worse, something of such magnitude that society has within a few seconds been turned upside down and rattled to its core. No radio stations; no medical services; no trash pickups; no subways; no newspaper delivery—no newspapers at all—no heat; no elevators; no water from the tap; no cell phones; no police cars; no food; no help.

Streets are clogged; the subway isn't running. Grocery and convenience stores quickly sell out of food and emergency supplies.

Financial systems fail—Wall Street, banks, ATMs—all are incapacitated. Smith & Wesson (NASDAQ: SWHC) replaces Apple (NASDAQ: AAPL) as America's most valuable company—except stocks aren't trading. Around the world, analysts and traders in financial institutions watch their computers in disbelief as the American economy drops offline. It's very clear that there will be no opening bell on Wall Street.

EMERGENCY ROOM HELL

The hospitals in every affected city are scenes of chaos and carnage, as backup generators and battery-powered systems that were not fatally damaged begin to falter and run down. Respirators, cardiac monitors, intravenous drip pumps, and dialysis machines all stop. Patients die, first by the scores, then by the hundreds, then by the thousands.

City leaders struggle to maintain some semblance of law and order, but they are quickly overwhelmed by the impossibility of controlling a world quickly spiraling out of control. Police are largely ineffectual. Groups of citizen angels form to fight roving gangs, but such efforts are quickly overwhelmed by the lack of communications devices. How can anyone call in a problem when there is nothing to call in with, or nothing with which to transmit the call on the other end?

By the thousands, people begin crossing New York City's bridges on foot, rucksacks over shoulders, in hopes of finding board at country farms or survivalist communes that had prepared for America's biggest vulnerability. But those hopes will soon be dashed by a view of the reality waiting over the distant hills.

Beyond the major city centers and their burgeoning suburbs and exurbs, the damage has been done in countless ways large and small. The blast over Pittsburgh has crippled virtually all agriculture and bulk transportation east of the Mississippi and north of Tennessee. The big farms, dependent on power for everything from milking machines to harvesting combines, are shut down. Powerless and darkened slaughterhouses and chicken processing plants are scenes of confusion—for people and animals alike.

Truckers are stranded in mid-run—truck stops, their gasoline and diesel pumps incapacitated, are islands of immobile vehicles of every size, and their fast-food joints have become no-food joints. Of interest to some drivers who are going nowhere is the number of cars and trucks that seem unaffected by whatever has befallen the majority of vehicles. A few older-model cars and trucks, and some motorcycles, are running—trying to negotiate the gauntlet of frozen sedans, vans, buses, and trucks scattered in their way.

The sprawling rail yards from Richmond to Chicago are silent. Miners deep in the earth have only their headlamps to guide them to lifts that no longer provide passage back to the surface. The massive ore carriers out on

Lake Superior—deprived of their steering and navigation electronics—are reduced to drifting in the wind and currents.

What has happened? Who is to blame? Could this have been prevented?

"What happened?" is an easy question to answer—at least from a physics perspective. A nuclear device capable of generating a massive electromagnetic pulse—an EMP of mega proportions compared to the tiny, man-held devices that paralyzed parts of Washington, or the larger truck-mounted versions that brought down the two airliners leaving Reagan National Airport—was detonated at a point one hundred miles over the eastern third of our continent.

The launching platform for the missile/warhead combination could have been a tried-and-true German-made submarine originally purchased by Venezuela's navy, but repurposed as the result of a political deal with a well-funded terrorist organization. (But let's be real: the sub could just as easily have been operated by India, or any one of our "friends" who decided to take an alternative route to screwing America.) All the *Sabalo*-type's crew needed to do was position the submarine close enough to the East Coast to get a missile with a basic nuclear warhead up and over a good portion of the U.S. and Canadian Eastern Seaboard.

This scenario is, in cold, hard fact, imminently real. That doesn't mean it would be easy. Our military's antimissile defenses are quite effective, as we'll discuss, and there's a very good chance the sub-launched missile would have been taken out. But there is no assurance of that. And there are many other ways to launch just as devastating an attack.

But let's trace back to the flash-bang of the nuclear warhead and the high-powered wave of particles it sent earthward as an electromagnetic pulse, an EMP, so we may better understand how it works.

As mentioned previously, an EMP is a high-intensity burst of electro-magnetic energy caused by rapid acceleration of charged particles. Nuclear weapons, nonnuclear weapons, and geomagnetic storms can power an EMP, and the resultant change in the magnetic field in the Earth's atmo-sphere can disrupt electrical systems.

An EMP has three main components. Scientists define them as: *E1*, *E2*, and *E3*.

- E1: An electromagnetic shock disrupts electronics, such as communication systems.

- E2: An effect similar to lightning rapidly follows and compounds the first component.

- E3: The pulse flows through electricity transmission lines, overloading and damaging transmission distribution centers, fuses, and power lines.

The most important mechanism for electromagnetic pulse production from a nuclear detonation is the ionization of air molecules by gamma rays generated from the explosion. These gamma rays interact with air molecules to produce positive ions and recoil electrons called *Compton electrons*. This pulse of energy, which produces a powerful electromagnetic field, is known as an *electromagnetic pulse*.

High-altitude nuclear detonations and electromagnetic bombs can generate EMP with the potential to damage or destroy every electronic device *within line of sight*. So the higher up in the sky the bomb is detonated, the further the destruction travels. Electrical systems within line of sight would be highly vulnerable. However, the EMP would not produce any serious damage outside the line-of-sight radius of the blast.

E1—LIGHTS OUT!

The E1 component of the pulse is the most commonly discussed element. The E1 pulse is produced when gamma radiation from a nuclear detonation knocks electrons out of atoms in the upper atmosphere. These electrons then spiral rapidly at 94 percent of the speed of light (or about 175,000 miles per second) around the Earth's lines of magnetic force, generating a powerful electromagnetic shock wave. This shock wave then spreads downward through the atmosphere along a line of sight from the detonation to the horizon. It covers a vast geographic region with an electromagnetic pulse that can insert thousands of volts into any metal object that can serve as an antenna—the metal bodies of aircraft in flight, automobiles on the ground, power and telecommunications lines, or even the cord on a personal computer.

Although the electromagnetic field created from an EMP lasts for only a short time, its effects can be devastating. The blast in our scenario over Pennsylvania was sufficient to send a destructive pulse deep into Canada, and as far south as Raleigh, North Carolina. Scientists have theorized that a single high-altitude burst two hundred miles above Kansas could spread an EMP pulse across the entire continental United States.

E2—MAKES LIGHTNING LOOK LIKE A WUSS

The E2 component is produced by weapon neutrons and is considered an intermediate time pulse, analogous to lightning. It lasts for up to a second after the beginning of the electromagnetic pulse. This E2 component has many similarities to pulses produced from lightning, and can be protected by breakers and surge arrestors used for defense against lightning. However, since lightning protection must be maintained in order to be effective, and since many commercial facilities and residences often neglect lightning protection, E2 can also be a serious threat. If an electronic component is protected *only* from an E2 pulse, an E1 pulse will destroy it.

E3—FRYING TRANSFORMERS

Rounding out the terrible triumvirate of EMP components is E3, or *magnetohydrodynamic electromagnetic pulse*. It is a relatively slower pulse and has similarities to a geomagnetic storm caused by what is referred to as a *coronal mass ejection*, or CME, a technical term for a massive burst of solar wind stemming from solar activity on the sun's surface (we'll get to that later). Like a geomagnetic storm, E3 can produce geomagnetically induced currents in long electrical conductors, which can then damage components such as power line transformers.[1]

3

OUR NATION'S CORE VULNERABILITY—THE POWER GRID

THE ELECTRICITY WE ENJOY every second of every day comes with a price. That price is making sure the power grid that feeds our cities, neighborhoods, schools, emergency responders, grocers, and the entire supply chain is always up and running. And that requires our immediate attention if we have any hope of continuing forward into the twenty-first century, and not backward into the early nineteenth century, due to an EMP attack.

Let's look at that grid.

The power industry—from the massive hydroelectric dams and their generators, to the hundreds of thousands of miles of power lines that stretch across the continent, to the local substations, to the transformers that hang on the pole down the street from your house, to the breaker box in your basement—is the foundation of our American civilization. Our lifestyle would quickly devolve to Third World drudgery without power and a method for transmitting it over long distances. Our survival is even more tenuous than in Third World cultures, because few of us possess any survival skills anymore. Unlike many Third World countries, we rely heavily on electronics to do the heavy lifting in our daily lives.

A NET TO SAVE US?

Our continental electrical grid is divided into two major and two lesser geographic zones, and then further divided into regional service areas and a patchwork of public, private, and cooperative utility companies. But these two large zones are each dependent, so a cascading power failure in one could shut down electrical service to at least half the nation and possibly the entire nation.

It is a tribute to our nation's power companies that we seldom experience brownouts, blackouts, or systemic failures, doubly so because with each passing year we place new and more challenging and complex demands on the grid. These potential "fail points" are many and varied:

- Electronic parts are exceedingly complex and interdependent, increasing the chances of one failure triggering another.

- Systems all across the country are often operating beyond capacity.

- Newer plants are being built farther away from end users, creating miles of exposed transmission lines.

- Newer energy sources, such as wind and solar, are being added to the mix, but these new sources are far less predictable.

- Large sections of the transmission system are not profit generating, so they tend to be neglected, unkempt, and vulnerable.

- Vital electronic systems are "remotely managed" and thus more at risk.

- Even without a power grid disaster, the big electrical transformers are made overseas and have a minimum one- to three-year production backlog.

- Newer transformers rely on single phase instead of triple phase (for ease of shipping) and these are more vulnerable.

- We have little in the way of energy storage, so in a disaster we could not rely on "reserves."

DON'T LOOK AT THE GRID TOO CLOSELY; IT MIGHT BREAK

It's something of a miracle that the power grid works as well as it does. It has been able to grow to accommodate swelling demands on it—thanks to giant advances in controls technology and operating practices. It has been able to "absorb" the consumption growth on existing power lines, with the addition of only a few new substations. However, there has been little construction of new transmission capacity, particularly the new, higher-gigahertz transmission lines required of the Internet age.

So the grid is effectively maxed, as they say. It is far less able to compensate for any potential difficulties than in the past. And the liabilities are only increasing with each passing day. The system is being stressed beyond reasonable limits. The electrical power system has become virtually fully dependent upon electronic systems working nearly flawlessly. The overall system reliability is testimony to the skill and effectiveness of the control systems. However, the very thin "margin for error" results in a greater likelihood of a catastrophic cascading outage. And should these systems be disrupted, even in a small way, the power grid could easily fail on a broad scale.

The reason for such a debilitating outcome is due to a national network of interlocking and interdependent critical infrastructures that depend entirely on electricity and electronics in order to function.

These critical national infrastructures require their harmonious function to allow for the myriad of actions, transactions, and information flow that undergird how our country's entire civil and military sectors function. In the case of our national grid system, for example, it already is in a state of needed repair in which it hasn't been tended to in years to update equipment or harden it against an EMP event.

Consequently, either a man-made or natural EMP attack on our national grid infrastructure or any of our other major infrastructures such as our telecommunications, transportation, petroleum and natural gas, finance and banking (including ATM) systems, food and water distribution, medical care, trade and production, emergency services and space systems could result in unprecedented cascading and devastating failures due to their interdependent relationship.

As Dr. William Graham who chaired the 2008 EMP Commission

pointed out, any recovery of any one of the key national infrastructures "is dependent upon the recovery of others. The longer the outage, the more problematic and uncertain the recovery will be.

"It is possible," he added, "for the functional outages to become mutually reinforcing until at some point the degradation of infrastructures could have irreversible effects on the country's ability to support its population."

Such a recovery, he added, could take months or years.

"The fact that key components of the U.S. electrical grid are not even manufactured in America and must be ordered a year in advance from foreign suppliers suggests just how complicated and time-consuming recovery might be," he said.

"The high state of automation within America's utilities further complicates recovery," he added. "There just might not be sufficient trained manpower available to get the job done in a timely way."

To add to this misery index, Graham said that the federal government does not today have sufficient human or physical assets for reliably assessing and managing EMP threats.

For example, an electromagnetic pulse of just 10 kilovolts per meter (kV/m), which isn't much if you've got a good flux gun at the ready, can cause an electrical charge a billion times more powerful than our ordinary electrical systems were designed to tolerate. The pulse occurs so quickly that the damage is done even before automated emergency shutdown systems are triggered. In some cases the pulse actually melts critical components. Wires in computer chips can be permanently damaged by a few tenths of an amp. An EMP with a field level of 100 kV/m can damage even unplugged electronics. The extent of the damage could cascade into 70 percent of the nation's power grid.

HOW PROTECTED IS THE POWER GRID FROM EMP?

Concerns about EMP attacks did not sprout up recently with the advent of the war on terrorism. Not at all. These concerns date back to the earliest days of the Cold War, when we first learned about the physics of it all. However, once America prevailed in that war with the Soviet Union and consigned them to the dustbin of history, U.S. defense strategists relaxed and stopped thinking about EMP. Until a few years ago. According to

Patrick Chisholm, writing in *Military Information Technology*: "A 2004 report by a panel of experts warned terrorists or other adversaries could launch an EMP attack without having a high level of sophistication, such as through short-range SCUD missile(s). Also of concern are nonnuclear, small-scale E-Bombs that target localized areas."[1]

Now, when you consider that we were once very concerned about an EMP assault, you might conclude that we have a number of defense systems in place—shields and defenses we could now fall back on. That would certainly be the logical conclusion. Our military has spent trillions of dollars since the Cold War began—surely we've hardened against such a *likely* attack scenario?

No. Even a cursory analysis shows that our computerized, electronically dependent society offers any rogue nation or its proxies a perfect target: an EMP-vulnerable power grid susceptible to a sucker punch to the heart of our infrastructure. Says Rep. Ed Markey of Massachusetts: "The electric grid's vulnerability to cyber and to other attacks is one of the single greatest threats to our national security."[2]

Representative Markey has introduced legislation to fund aggressive defenses against EMP attack. He points to events such as the August 2003 blackout, which struck from Michigan to Massachusetts, as a warning knell. There were minimal deaths and injuries tied to the 2003 blackout; only a few died in carbon monoxide poisonings as a result of generators running in their homes or from fires started from candles. But the effects were pervasive. Television and radio stations went off the air in Detroit; traffic lights and train lines stopped running in New York, turning Manhattan into the world's largest pedestrian mall; and water had to be boiled after water mains lost pressure in Cleveland.[3]

In widespread power outages of the past, people reacted with behavior ranging from rioting and looting (as many did during the 1977 New York outage) to patiently waiting for the crisis to be over (as with the 2003 outage). But if the recovery period were longer, and if electronic communications were down for a period of weeks or even months, civilization in the United States could reach a tipping point, where people's worst impulses would be on display and recovery would be very difficult.

THE PULSE THAT CAN STOP OUR NATION'S HEART

Over the past two decades, very few large-capacity electric transmission systems have been constructed in the United States. And most of the new equipment has been purposefully located at great distances from the customers it serves. This has been done in an effort to mitigate both environmental and economic concerns. But it has added stress to the transmission systems and further limited the systems' ability to withstand disruption.

Significant elements of the systems, including many generating plants, are aging and becoming less reliable or are under pressure to be retired for environmental concerns, further exacerbating the situation. In fact, a considerable number of these electric transmission systems are more than fifty years old.[4]

The transformers that handle electrical power within the electrical transmission system are large, expensive, and often custom-built. Given how they often are customized for local area use, this will pose a serious problem if they are knocked out. An E1 electromagnetic pulse would arc across the insulators that separate power lines from their wooden or metal pole supports. The arcing would damage the insulators themselves and, in some cases, can cause transformers mounted on the poles to explode. Damage to large numbers of insulators and pole-mounted transformers would also result in a shortage of replacement parts, since they are considered sturdy and reliable items under normal conditions. Consequently, spares are not kept to cover widespread losses, resulting in delays in finding replacement parts. Replacing them on a timely basis will prove difficult, if not impossible.

There are currently two thousand large transformers rated at or above 345 kV in the United States alone, and about twenty of them are replaced every year. Worldwide production capacity is less than one hundred units per year and serves a world market—one that is growing at a rapid rate in such countries as China and India.

Delivery of a new large transformer ordered today is nearly three years, including both manufacturing and transportation. An event damaging several of these transformers at once means it may extend the delivery times to well beyond current time frames.

How much longer would it take to replace blown-out transformers if hundreds or even thousands were suddenly destroyed and needed to be replaced?

POWER GENERATION PLANTS ALSO AT RISK

Like transmission systems, power generation plants are vulnerable to EMP. These generation plants have been built to withstand blackouts while their protective systems generally take over to run again. However, an EMP can upset the protection and control system, damage critical components, and trigger an emergency shutdown.

There are some five thousand generating plants of significance nationwide that will need some form of protection against an electromagnetic pulse, particularly their control systems. Power plants, particularly newer ones, are highly sophisticated, very high-speed machines, and improper shutdown can damage or destroy many critical components and can even cause a catastrophic failure.

Generally, a proper shutdown is orderly and depends on a synchronized operation of multiple controllers and switches. For example, cooling systems need to respond properly or temperature changes during shutdown can cause a boiler rupture. The turbines require an orderly spindown to avoid shaft sagging and blades from hitting the casings. Bearings can easily fail and freeze, damaging the shaft if the shutdown doesn't get lubrication. As with any complex machine operating at high temperatures at fast speeds with tight tolerances, the power plant survivability will depend on a large number of protective systems, creating higher prospects for plant damage and failure.

SCADAS: LITTLE BOXES THAT RUN OUR LIVES

You could easily get through life never hearing the acronym *SCADA*, yet you are exposed to its silent magic thousands of times a day. SCADA is short for *supervisory control and data acquisition*—a fancy term for industrial control systems. Their electronic switches run everything from streetlights to railroad crossing gates to irrigation systems to the generators that power our cities.

The U.S. power industry alone invests an estimated $1.5 billion a year in new SCADA equipment. You'll find these apparatuses in electrical transmission and distribution systems, water management, oil and gas pipelines, utilities and telecommunications, nuclear power plants, and just about everything else we commonly call *infrastructure*—certainly one of our more boring terms. But if this infrastructure, with all of its SCADAs, were ever struck by an EMP attack, our lives would quickly become a living hell.

SCADA oversees all operations free of human hands, making its own electronic adjustments based on complex built-in algorithms. If an electrical generating plant, for example, fails due to the loss of a critical hardware component or industrial accident, the SCADA will detect the loss and issue an alert. In turn, the SCADA will issue commands to other generating plants under its control to increase power output to make up for the loss. (There are major interdependencies among all these infrastructures. For instance, a phone company cannot operate without the electricity company.) All of these actions occur automatically, within seconds and without human involvement.

SCADAs are often located in remote areas, so they are exposed, with unprotected cables and easy fault points. An EMP attack would most certainly cause serious physical damage to the SCADA controls, and in turn, create the very outages SCADA was intended to prevent.

SCADA was considered so important that the 2008 EMP Commission singled out these "ubiquitous robots of the modern age" as "job one" in protecting the homeland. This EMP Commission, which you'll hear a lot more about in these pages, studied every aspect of the EMP threat in hopes of warning Americans and preparing us for the warfare of the future. About SCADA in particular, the EMP Commission had this to say: "While conferring economic benefit and enormous new operational agility, the growing dependence of our infrastructures on these omnipresent control systems represents a new vector of vulnerability in the evolving digital age."[5]

This problem of advancing technology and interdependence became apparent in 1991 when a single fiber-optic cable was severed just outside of New York City. A full 60 percent of all calls into and out of New York went dead. All air traffic control functions between Washington, D.C., and Boston (the busiest flight corridor in the nation) were disabled. Operations of the New York Mercantile Exchange were halted. Yet nobody could do a thing about it. Nobody had a contingency plan for such a situation. No attempts had been made to develop operational workarounds.

UNLESS YOU SHUT DOWN THE CITY, YOU CAN'T TEST SCADA

These electronic controls are now so pervasive, so central to the provision of all the modern amenities we take for granted, that we assume they'll keep working—no matter what. And we *have* to assume this, because we really cannot adequately test the SCADAs . . . without shutting down and blacking out entire geographic regions. Nobody wants that.

So we do not test SCADAs with any rigor, other than with hypothetical simulations and models. We really don't know what could happen, or rather, what will happen. The EMP Commission, with all of its congressional funding and resources, tried in earnest to simulate the effects of an EMP attack, but admitted, "To provide insight into the potential impact of these EMP-induced electronic system malfunctions, one can consider the details of historical events. In these cases, similar . . . system malfunctions have produced consequences in situations that are far too complex to predict beforehand using a model or analysis."[6]

Even simulated testing isn't likely to tell us how drastic a multiple-SCADA hit would be. Our nation has created such a complex of interconnections across so many millions of vital moving parts that anything could happen. We simply "cannot possibly anticipate all the possible interactions of the inevitable failures," reminds Charles Perrow in *Normal Accidents: Living with High Risk Technologies.*[7]

Perrow's thesis is both academic and fascinating: Our attempts to build more and more complex mechanisms make failure more and more inevitable. These failures become an inherent expectation of a tightly coupled system that now pushes the thresholds of its original design.

THE COST OF REBUILDING AFTER AN EMP ATTACK

We've focused thus far on the devastating effects of an EMP attack on electronics. But what would be the costs borne by national, state, and local economies?

To answer this question, the Sage Policy Group of Baltimore released a study in 2007 attempting to make a "conservative determination" of economic impacts. They focused on the Washington, D.C., region—from Baltimore to Richmond. The goal was to "put a financial face to the problem and suggest quick steps that can be taken to mitigate it" following an EMP attack.[8] The study, commissioned by Instant Access Networks, LLC, looked at the economic effects of a nuclear device detonated at high altitude. Consider these key excerpts: "A larger nuclear device detonated at 300–400 miles above ground would impact the entire continent and may also produce a substantial slower pulse (this is the E3 component) that can cause additional damage . . . In these instances of high-altitude EMP, no one would feel the heat or blast but merely experience the effects of the disruption or damage to the electronic and power infrastructure."[9]

The study estimated that the effects of the detonation would extend out hundreds of miles beyond the Washington region, "significantly complicating the recovery process and the restoration of economic activity while producing economic consequences roughly ten times greater than those impacting" only the Washington region.[10]

The Sage report presented a range of low-, medium-, and high-impact scenarios and the resulting estimates of the economic damage. In the worst-case scenario:

- The damage from an EMP attack would be widespread, and the duration of disrepair would last years.

- The quantity of equipment needed to replace what was destroyed and damaged would quickly exhaust readily available supplies and could exceed existing manufacturing capacity.

- The available skilled labor to replace and restore key infrastructure probably wouldn't exist for some time.

- Replacement and repair would occur slowly and only gather speed as the basic infrastructure was gradually brought back online.

- The impact to financial output over a thirty-three-month period ranges between $100 billion and $300 billion.

But if preparedness measures were taken beforehand, says Charles Manto, who commissioned the study, the savings could be substantial: "The three city metro area between Richmond, VA, Washington, D.C. and Baltimore, M.D., could save $25 [billion] to $185 [billion] in losses if they were to take steps to protect their most critical communications and power infrastructure. The East Coast as a whole would save ten times as much. . . extrapolating to the East Coast area as a whole would mean a ten times larger loss of one to three trillion dollars."[11]

Limiting the risk from the geomagnetic storm–type threat involves stockpiling large transformers and installing dampers, essentially lightning rods, to dump surges into the ground from the grid. Even if such steps cost billions, the numbers come out looking reasonable compared with the $119 billion that a 2005 Electric Power Research Institute report estimated was the total nationwide cost of normal blackouts every year.[12]

REAL-WORLD ACCIDENTS TELL THE STORY

While we cannot possibly anticipate all the paths to failure, we can look at real-world accidents to begin measuring the magnitude of destruction we would see from widespread SCADA failure.

NETHERLANDS, 1980s
In the 1980s, there was a large explosion in a thirty-six-inch natural gas pipeline in the Netherlands, triggered by a nearby naval radar installation. The radio frequency from the radar caused the SCADA to rapidly open and close a control valve, creating pressure waves that traveled down the pipeline and eventually caused an explosion. It took authorities six weeks to determine the cause.

UNITED STATES, 1990s
In Bellingham, Washington, in the late 1990s, there was a similar SCADA malfunction. Some 250,000 gallons of gasoline from a pipeline were dumped into the Hannah and Whatcom creeks. The fuel in the water ignited, killing three people and injuring eight. Explosions collapsed creek banks for a mile and a half, damaging many buildings near the water. In the aftermath, it was discovered that the culprit was a lethal combination

of inadequate maintenance and improperly managed SCADA.

UNITED KINGDOM, 1994
A severe thunderstorm struck the Pembroke oil refinery, creating a half-second power loss and power dip. Numerous pumps and coolers tripped offline. An explosion followed. As a result, an estimated 10 percent of the UK's refining capacity was lost until the repairs were made. Business losses amounted to $70 million—all from a single lighting strike, which is, after all, the closest thing to an EMP attack.[13]

UNITED STATES, 1999
San Diego County Water Authority and San Diego Gas and Electric both experienced unidentifiable electromagnetic interference to their SCADA wireless networks. They were unable to open or close critical valves as a result. Technicians were dispatched to remote locations to manually open and close the water and gas valves. That saved a "catastrophic failure" of their water system according to a review commission.[14] Had there been a catastrophic failure, as much as 825 million gallons of water a day would have flooded the city. Investigators later determined that this SCADA failure had been triggered by the radar of a ship sitting twenty-five miles off the coast.

UNITED STATES, 2000
A pipeline near Carlsbad, New Mexico, operated by the El Paso Natural Gas Company, exploded, killing twelve and leaving an eighty-six-foot crater. The explosion occurred due to maintenance failures triggered by an electromagnetic pulse event.

There have been many more accidents tied to SCADA failures, but these illustrate just how vulnerable we really are to an EMP attack, and how widespread the damage could quickly become. As the EMP Commission concluded, "It is not one or a few SCADA systems that [would be] malfunctioning . . . but large numbers—hundreds or even thousands—with some fraction of those rendered permanently inoperable until replaced or physically repaired."[15]

Nobody has planned for a scenario with the loss of hundreds or even thousands of nodes across all of the critical national infrastructures, all at once. That, however, is precisely how an adversary could seek to attack and debilitate us.

LOSING THE NATURAL RESOURCES WE VALUE MOST

Petroleum and natural gas. We rely on them for everything. To power our cars, heat our homes, run our factories, and be turned into a thousand refined and extruded products, from fertilizers to plastic toys. To keep these vital resources moving toward end users, we've built an extensive national pipeline system.

We have 180,000 miles of interstate natural gas pipelines; 300,000 miles of interstate and intrastate natural gas transmission lines; another 1.8 million miles of smaller distribution lines that carry natural gas right to homes and businesses; and a nationwide network of production wells, processing stations, and storage facilities. To transport oil we have 55,000 miles of large-diameter oil pipelines, or trunk lines, along with some 40,000 miles of smaller-diameter pipelines, or gathering lines. Together these lines spider web the country to comprise the national delivery system. In addition, there are 95,000 refined product pipelines that bring these products to their final destination, along with trucks that are used to transport the refined products of gasoline, diesel, and jet fuel to the marketplace.

The entire petroleum and natural gas delivery system relies on SCADA and is thus susceptible to dangerous malfunctions that could lead to massive fires and explosions as a result of an EMP event.

For natural gas, the backup emergency pressure regulators make unlikely a failure that would cause any rupture or explosion. (A little good news, finally.) The most likely result, if there is no manual intervention, would be a significant loss of pressure after a period of time, leading to a substantial disruption of service.

With the oil, if the SCADA system no longer worked, then operations would have to be shut down. A petroleum pipeline failure could be catastrophic. Leaking oil could contaminate water supplies and cause disastrous fires.

In a refinery, which is dependent on computers and integrated circuitry in process control, a failure would trigger an emergency override to allow for a controlled shutdown of the refinery. However, a fast shutdown could seriously damage the refining equipment and force the system off-line for an extended period, at tremendous cost to end users.

The BP oil derrick explosion of 2010 is a stark reminder of what happens from equipment failure. The huge oil spillage affected five states and caused a major loss of employment and income from which the region is still struggling to recoup.

So where does that leave the United States with respect to other nations (not all of whom are our friends) who have gained EMP knowledge and who may be itching to couple it with their improving missile technology?

EMP AS A WEAPON?

We've now looked at what could happen in the milliseconds following an EMP detonation, and what it might cost us if we're not prepared to take that hit. But the questions we should answer first are these:

1. Have we truly proven that these EMP scenarios could happen?
2. Do we have evidence that an EMP attack is being planned?

The answer to both questions is, "Yes!"

FROM A DESERT TO A FISHBOWL

The United States first became aware of the "electromagnetic pulse" when we detonated the world's first nuclear device on July 16, 1945, in the desert sands of Alamogordo, New Mexico. From that moment on, under wartime conditions at Hiroshima and Nagasaki, and through decades of above- and below-ground tests, the effect of nuclear-generated electromagnetics came under increasing scrutiny.

In July 1962, the United States ran a series of high-altitude tests, code-named Operation Fishbowl. These tests significantly advanced the knowledge of the EMP effect. One of these tests in particular—known as Starfish Prime—was conducted at a height of 250 miles above the Pacific Ocean.

A 1.4-megaton thermonuclear warhead, when detonated so high up in the sky, made no sound. There was a very brief and very bright white flash seen—witnesses likened it to a giant flashbulb going off. The flash could be seen even through the overcast skies at Kwajalein Island, 1,200 miles away. After the flash, the entire sky glowed green over the mid-Pacific for a second; then a bright-red glow formed at "sky zero," where the detonation had occurred.

Long-range radio communications were disrupted for a period ranging from a few minutes to several hours (depending on the frequency and the radio path being used).

Do we have evidence that any such attack is being planned?

There was significant electrical damage in Hawaii, almost nine hundred miles from the site of the detonation. The blast knocked out streetlights, set off burglar alarms, and damaged the local telephone company's microwave link, according to a report by Charles N. Vittitoe of Sandia National Laboratories, in a June 1989 article titled "Did High-Altitude EMP Cause the Hawaiian Streetlight Incident?"

The radiation cloud from the Starfish Prime test *subsequently* destroyed at least five U.S. satellites and one Soviet satellite. *Telstar I,* the world's first active communications satellite, was damaged by the radiation cloud and failed completely a few months later.

Starfish Prime was followed by two other high-altitude tests in November 1962—Bluegill Triple Prime and Kingfish. These tests provided enough electromagnetic pulse data for scientists to accurately identify the physical properties producing the pulses.

From these tests, a volume of findings emerged. There was little in the way of serious damage in the South Pacific—but then, there was little in the way of modern development in the blast area. That was the point! However, if the same explosion had occurred over the northern portion of the continental United States, U.S. scientists concluded, the damage would have been quite significant. This is also due to the greater strength of the Earth's magnetic field over the United States.

As more tests were conducted, the global scientific community was brought in and the world began to gain a fuller understanding of EMP and its effects. These effects were best articulated in a series of three 1981 articles in *Science* by journalist William J. Broad. An excerpt:

The United States is frequently crossed by picture-taking Cosmos series satellites that orbit at a height of 200 to 450 kilometers above the earth. Just one of these satellites, carrying a few pounds of enriched plutonium instead of a camera, might touch off instant coast-to-coast pandemonium: the U.S. power grid going out, all electrical appliances without a separate power supply (televisions, radios, computers, traffic lights) shutting down, commercial telephone lines going dead, special military channels barely working or quickly going silent.[16]

A FAILURE OF IMAGINATION . . . AGAIN?

It has often been said that 9/11 was a failure of our imagination. The fact is, we had warnings about terrorists flying airplanes into buildings, but we chose to ignore them. We decided to not go there. Indeed, psychologists tell us that we subconsciously disassociate from certain things—we move them out beyond our waking imaginations. It's a form of self-preservation in an uncertain world. We just don't think about it.

The EMP threat falls into this category of denial. Our government fails to imagine it possible, even refuses to imagine it, despite overwhelming evidence that it can and will happen to one degree or other. In that very real sense, the government is failing in its *primary* task of protecting the citizens.

On the seventieth anniversary of the Japanese attack on Pearl Harbor, Warren Kozak wrote in the *Wall Street Journal* that the skeptics of EMP might be right. We may never experience a serious EMP attack on native soil. But, he also noted, "70 years ago, similar doubters believed Japan would never be so foolish as to take on the United States of America—until, of course, it did."[17]

And the Pearl Harbor sneak attack was for the Japanese a stupendous undertaking, with logistics and planning far surpassing what would be required to pull off an EMP attack—an attack that the members of the EMP Commission concluded is right now being contemplated by hostile nations. Research done for the commission by staff member Dr. Peter Vincent Pry found that "the physics of EMP phenomenon and the military potential of EMP attack are widely understood in the international community, as reflected in official and unofficial writings and statements."[18]

Dr. Pry is an authority on EMP and defense policy. He has testified that the countries possessing knowledge of EMP weaponry include Britain,

France, Germany, Israel, Egypt, Taiwan, Sweden, Cuba, India, Pakistan, Iraq, Iran, North Korea, China, and Russia. Dr. Pry told the U.S. Senate Judiciary Committee's Subcommittee on Terrorism, Technology, and Homeland Security that these nations feel obligated to develop EMP weaponry because the U.S. is doing so. According to Dr. Pry, "Many foreign analysts—particularly in Iran, North Korea, China, and Russia—view the United States as a potential aggressor that would be willing to use its entire panoply of weapons, including nuclear weapons, in a first strike. They perceive the United States as having contingency plans to make a nuclear EMP attack, and as being willing to execute those plans under a broad range of circumstances."[19]

So these nations are conducting tests of their own . . .

IRANIANS LAUNCHING SCUDS FROM SHIPS?

U.S. intelligence has long monitored Iranian ship communications, and one particular intercept in 1998 proved deeply troubling.

Apparently, a lone cargo ship had slowed to a stop in the Caspian Sea off the coast of Iran. Feverish activity was under way on the deck. It was night, and only a few lights aided the sailors as they heaved away the tarp covering a large structure on deck. Then came the sound of a hoist from a "TEL" or transporter erector launcher, raising a grayish missile to a vertical firing position.

A sudden burst of exhaust illuminated the night as the missile slowly rose and began to tilt northward. U.S. naval telemetry monitored the missile as it traveled two hundred miles before splashing into the Caspian near Russia's coast. U.S. intel also picked up Iranian ships anchored within sight of the splashdown area, doing their own monitoring and assessment of the operation.

Further U.S. intelligence analysis of the missile's signature verified that it indeed was a Scud that had been fired from an Iranian ship under control of the Army of the Guardians of the Islamic Revolution. This group also controlled Iran's fighter aircraft and its strategic missile forces, which included several thousand short- and medium-range missiles.

At the time, those missiles were crude and undependable—though obviously improving enough to hit a target two hundred miles away.

Today, the improvements have been substantial. Iran's modern missiles include the Shahab-3 series, which gives the Islamic republic intercontinental ballistic and space capability.

Iran, however, is not the only country either developing or having an EMP capability to strike targets worldwide.

RUSSIA DESPERATELY WANTS TO REGAIN A SEAT AT THE TABLE

Analysts at the Congressional Research Service (CRS) have detailed Russia's efforts to protect its infrastructure against an EMP attack by hardening both civilian and military electronics. Several of CRS's sources claim the Russians have their leading physicists focusing on EMP, and there's no reason to think otherwise—given Russia's history of "firsts" in military science.[20]

In October 1962, the Soviets launched a warhead on an R-12 missile to test the effects of EMP on military systems. They would later report their shock at the magnitude of damage inflicted. In the test, known simply as Test 184, the Soviets detonated a 300-kiloton thermonuclear warhead at an altitude of 180 miles over central Kazakhstan. It knocked out a major underground power line for 600 miles. Fires were reported in several electrical power plants. An overhead telephone line was fried for 350 miles of its length when current surges of 1,500 to 3,400 amperes ran through the lines. Radios sitting 350 miles away were knocked out. Military diesel generators were also damaged. That was the most surprising aspect in the scientists' view, since the generators rely on mechanics over electronics, and generators would necessarily become very important following an EMP attack.[21]

Both the Soviet Union and the United States detonated EMP-generating nuclear weapons tests in space during the darkest days of the Cuban missile crisis, when the world was on the brink of nuclear war.

The 2008 EMP Commission also says the Russian military claims to have developed a super-EMP weapon capable of generating 200 kilovolts per meter. This claim means that a conductive object exposed to an electromagnetic pulse would receive a surge of 200,000 volts for every meter of length. Given that most electronics run on a few volts, the overkill implied here is enormous.

CHINA ENTERING THE EMP RACE EYES WIDE OPEN

In his testimony to Congress, Dr. Pry outlined the efforts by the Russians and Chinese as well to turn EMP into the primary or even the *only* means of attack in a wartime scenario.[22] Additionally, July 2008 testimony by then-assistant secretary of defense for Asian and Pacific Security Affairs, James J. Shinn, told of detailed Chinese plans to threaten Taiwan.

A recently declassified 2005 report from the Army's National Ground Intelligence Center (NGIC) outlined Chinese efforts to develop both EMP and high-powered microwave weapons.[23] The NGIC said the Chinese could disable electronics on U.S. aircraft and warships: "China's [high-altitude] EMP capability could be used in two different ways: as a surprise measure after China's initial strike against Taiwan and other U.S. [aircraft carrier strike group] assets have moved into a vulnerable position, and as a bluff intended to dissuade the United States from defending Taiwan."[24]

That Army intelligence report also outlined specific strategies mainland China could use in attacking Taiwan with an EMP. The report further indicated the thoroughness of the Chinese in this realm; they have been testing animals for EMP effects:

> Chinese medical researchers presented three briefings at the Asia-Pacific Electromagnetic Fields, Research, Health Effect, and Standards Harmonization Conference in Bangkok, Thailand (26 to 30 January 2004), on the bio-effects of Intense HPM and EMP radiation. Although the data presented related only to animal experiments (mice, rats, rabbits, dogs and monkeys), all three briefings made it clear that the real purpose was to investigate potential human effects of exposure to these specific radiations. One brief, "Effects and Mechanisms of EMP and HPM on Optical Systems in Monkey, Dog, and Rabbit," dealt mostly with eye injury. "Bio-effects of S-Frequency High Power Microwave Exposure on Rat Hippocampus" dealt mostly with brain injury. The third and final briefing, "The Species Specificity and Sensitive Target Organs of Injury Induced by Electromagnetic Radiation (BIO-EFFECTS OF EMP AND HPM)," dealt with species-related injury thresholds for all affected organs. Dose-effects relationships were established in all three studies . . .
>
> By a process of extrapolation, one might be able to use these data to set danger thresholds for human exposure. The high mortality rates of animals (especially for primates) exposed to EMP radiation in

the recent Chinese experiments are in graphic contrast to the lack of reported bio-effects associated with EMP exposures during the period of atmospheric nuclear testing (1950s/1960s) by the United States and other nations. This is probably a consequence of the extremely high field strengths used in the Chinese experiments.[25]

From even a layman's reading of these unclassified passages, it appears that China is developing EMP weapons to be able to defeat a technologically superior U.S. military in battle, should it come to that.

INDIA—YES, INDIA—COULD BECOME THE BIGGEST THREAT

In *The Shoes of the Fisherman*, the 1963 novel by Morris West, China is in peril as food shortages—resulting from a U.S.-led trade embargo—threaten to bring the nation to its knees. Chinese leaders prepare for war with the Soviets to expand their geopolitical holdings and protect Chinese interests. Such a war, it is feared, could quickly escalate into a global conflagration. At the last minute, the Russian-born pope steps in and commits the Catholic Church in support of China. Tensions ease; war is averted; the Chinese get food.

Fast-forward fifty years, and look at India. It's a nation surrounded by countries that don't like it and don't trust it. Yes, the brainpower of the Indian people is impressive; their call centers are maddeningly efficient; and their economy is—well, venture twenty feet outside the Mumbai city limits and you're stuck back in nineteenth-century squalor and hopelessness. People are starving and solutions are nonexistent. Through India's great modernization, the poor have only increased in numbers. And that is the problem.

India's national pain is very real and not lost on leaders in India who control a nuclear arsenal. Unlike Iran or North Korea, India has fully operational nukes. And they know how to deliver them, if they need to. The question is, will they ever *think* they need to? If India decides they have to push out beyond their borders, to annex needed farmland or to take up arms against the hated Pakistanis, they might keep going—pushing into China for additional territory, even at the high price that would exact.

Russia, for its part, would sit by and watch—thrilled. They could cut a sweet deal with India to help it succeed.

This is a low-probability scenario, of course, but it is not at all inconceivable.

The Indian military has studied EMP in detail because it fears that Pakistan might use EMP against Bangalore, their version of Silicon Valley. And India has successfully test-fired long-range ballistic missiles capable of carrying a nuclear warhead as far as Shanghai. India's Agni-V is capable of carrying *MIRVs*, or *m*ultiple, *i*ndependently targetable *r*eentry *v*ehicles, each its own collection of separate warheads.[26] The missile also puts India closer to being able to develop antisatellite weapons, and the Agni-V appears to be capable of being launched from mobile platforms. All of this makes the missile a fearsome deterrent against foreign attacks. India will eventually be able to turn the Agni-V into an intercontinental missile capable of reaching Europe and the United States.

THIS IS A HARBINGER OF MISSILE PROLIFERATION TO COME

With all that's going on, the U.S. State Department seems unconcerned, even labeling India one of our closest allies. The smug and comfortable bureaucrats of Foggy Bottom ought to be reminded, perhaps, that the United States has never slept peacefully with our *friends*. The British came back after the Revolutionary War to burn Washington. The Russians took a big bite out of Berlin after we helped them in World War II. And Iran turned on us in a heartbeat once the shah was gone. In this day and age of global mistrust and high economic stakes, it is not inconceivable that even an American ally like India—or even its Pakistani adversary—could undergo a regime change that would place our heartland in the crosshairs.

PAKISTAN, THE NEW AFGHANISTAN, WITH RUNNING WATER

Pakistan's relations with the United States are at the lowest ebb ever, in the wake of the unannounced raid in May 2011 on Osama bin Laden's compound. Pakistan has nuclear weapons and has said they will share those weapons with their close ally Saudi Arabia.

But the larger threat of Pakistan is the safe haven it offers dozens of terrorist groups. There are at least thirty-two international terrorist groups and four extremist organizations operating in Pakistan. This intel comes from the authority on the subject, the *South Asia Intelligence Review*, which provides weekly assessments and briefings on terrorism in the region. Following is a list of those groups:

INTERNATIONAL ORGANIZATIONS

1. Hizb-ul-Mujahideen (HM)
2. Harkat-ul-Ansar
3. Lashkar-e-Toiba (LeT)
4. Jaish-e-Mohammad Mujahideen E-Tanzeem (JeM)
5. Harkat-ul Mujahideen
6. Al Badr
7. Jamait-ul-Mujahideen (JuM)
8. Lashkar-e-Jabbar (LeJ)
9. Harkat-ul-Jehad-al-Islami (HUJI)
10. Muttahida Jehad Council (MJC)
11. Al Barq
12. Tehrik-ul-Mujahideen
13. Al Jehad
14. Jammu and Kashir National Liberation Army
15. People's League
16. Muslim Janbaz Force
17. Kashmir Jehad Force
18. Al Jehad Force
19. Al Umar Mujahideen
20. Mahaz-e-Azadi
21. Islami Jamaat-e-Tulba
22. Jammu Kashmir Students Liberation Front
23. Ikhwan-ul-Mujahideen
24. Islamic Students League
25. Tehrik-e-Hurriat-e-Kashmir
26. Tehrik-e-Nifaz-e-Fiqar Jafaria
27. Al Mustafa Liberation Fighters
28. Tehrik-e-Jehad-e-Islami
29. Muslim Mujahideen
30. Al Mujahid Force
31. Tehrik-e-Jehad
32. Islami Inquilabi Mahaz

EXTREMIST GROUPS
1. Al-Rashid Trust
2. Al-Akhtar Trust
3. Rabita Trust
4. Ummah Tamir-e-Nau

Many of these terrorist groups operate with a high degree of freedom in and from Pakistan. Given the increasingly rocky relationship between Pakistan and the United States, Pakistani terrorist groups will continue to plot against the U.S., and now many of them have a knowledge of or even could be holding EMP and RF weapons.

This is not an encouraging picture.

4

A DIRECT HIT—FROM THE SUN

O N A FARM NEAR SHELTON, NEBRASKA, about one hundred miles west of Lincoln, on the banks of the South Platte River, the morning sun begins its long climb up the early summer sky. Ninety-two million miles away, the swirling sphere of light and heat is to the Nebraska farm family what it has always been: the benevolent key to their very existence. Until today.

Above the farm, skimming the lower reaches of our atmosphere, thousands of commercial jets, private planes, and military aircraft crisscross the North American continent, bathed in the warm glow of Earth's nearest star. Somewhat closer to the sun—by about two hundred miles—the crew of the International Space Station awaits another shipment of supplies from the SpaceX robotic shuttle launched from Cape Canaveral two days earlier.

As the delicate ballet between the space station and the supply ship plays out, thousands of low-earth orbit and high-earth orbit satellites perform their mundane tasks: watching the weather; relaying voice and image signals; examining the intelligence operations of dozens of countries; monitoring the environment; and providing navigation signals to ships, planes, trains, cars, trucks, pedestrians, construction companies, surveyors, and more. Among the constellation of satellites, two—the Solar and Heliospheric Observatory and the Solar Dynamics Observatory—cast their unblinking eyes toward the sun, scrutinizing every aspect of its activity—taking its pulse, as it were—because once in a while, our star misbehaves. Today, both satellites, along with scientists in a sun-dedicated observatory in Arizona, witness the early signs of a major solar tantrum—and it is terrifying.

Boiling off the sun's surface during an intense geomagnetic storm, an enormous plasma flare a thousand times the size of Earth is sending a mass of charged particles toward our home planet at nearly the speed of light. It is not going to graze us; it will be a direct hit. And if it hits in just the right—or rather, wrong—alignment with our own magnetic field, it will become a G5 storm, the worst. The scientists alert the list of concerned parties worldwide, and a two-day wait begins.

If it turns out to be a low-level G1 storm, like the one in early 2012, damage will be minimal. But if not . . .

When the solar belch does slam into Earth's geomagnetic field, the nighttime skies over the northern reaches of the globe light up with auroral displays unlike any in memory. That will be the only positive report on this particular solar attack.

In an instant, literally, radio communications and other forms of telemetry become highly unstable. Worldwide phone calls fluctuate between fair to "Can-you-hear-me-now?" And satellites—many of which are not fully hardened or sufficiently protected against a direct hit by a G5 solar storm—either stop functioning, or malfunction, losing their orientation, spinning out of control, and careening into blue skies, where they burn to ash.

NOWHERE TO RUN, BARELY A PLACE TO HIDE

For the astronauts aboard the space station, the thin shielding of the main walls is not enough to keep them from being irradiated by the storm's brutal pulse. The crew rushes into the Soyuz module. With its thicker skin, it may protect them for the duration of the flare. However, the SpaceX cargo ship, now only a few thousand yards from the station, takes a direct hit from the pulse, and its systems go haywire, firing the robotic ship's thrusters and hurtling it on a collision course with the space station.

Closer to Earth, the pilots of dozens of flights are seeing confusing symbology on their flight deck displays. The navigation signals to and from the constellation of global positioning satellites are too far out of normal parameters to be believed. Airports that should be five hundred miles away are now shown as only two hundred miles out. The runway that should be right ahead of a plane emerging from the clouds on an instrument approach is almost two hundred yards to the right of the plane's course. Radio communications with air traffic control are either nonexistent or thoroughly garbled. Eventually, there is no contact at all between those in the sky and those on the ground, because the electrical power grid that supports the whole aviation and airways system has been hammered into silence by the sun's untimely—but not all-that-unusual—outburst.

Clearly this is not a G1 geomagnetic storm. It may even be a G5, and those airplanes should have been grounded. But no such decision was made; few protocols exist for such contingencies.

Back on the ground in Nebraska, the power grid first deteriorates and then fails entirely. The farm family kicks on the backup generator and doesn't miss a beat. If the cows have to be milked by hand, so be it. Down the road, the big corporate dairy struggles with the failure of all its milking machines. And the hands can be seen staring, rather stupidly, up at the unusually bright sun, uncertain what else to do.

When the power died all across the country, so, too, did all the hydroelectric systems—water pumps and generators, and all the specialized electronic switches (the SCADAs) that pretty much run everything that can be turned on or off. Now, thanks to the solar flare—a confirmed G5—everything electronic is no more.

Our farm family, ever resourceful, has got the old windmill running, and it is pumping water for the house, livestock, and vegetable garden. An older Ford F-150 pickup truck, its electromechanical ignition harness unaffected by the solar sword, runs just fine. And the farm's above-ground gasoline tank has plenty of fuel.

They may not have phone service, but that's no great loss. They enjoyed the cable TV, but oh well. The Scrabble game, a deck of playing cards, and a wall of books will keep the family entertained in the evening by beeswax candles—the ones they've been making for years.

This family, and its farm, will prevail.

HOW DO SCIENTISTS KNOW THIS COULD BE "THE BIG ONE"?

History has a way of telling us, if we pay attention to it. On September 2, 1859, the largest recorded geomagnetic storm dazzled the pre-electronics era with auroras seen around the world. The white-light solar flare was so bright, folks in the northeastern United States could read newspaper just from the light of the aurora. The telegraph system in use all over Europe and North America failed.

Telegraph operators were shocked, and telegraph pylons threw sparks, causing telegraph paper to spontaneously catch fire. And when telegraphers disconnected the batteries powering the lines, aurora-induced electric currents in the wires still allowed messages to be transmitted! According to the next day's *Baltimore American and Commercial Advertiser*:

> Those who happened to be out late on Thursday night had an opportunity of witnessing another magnificent display of the auroral lights. The phenomenon was very similar to the display on Sunday night, though at times the light was, if possible, more brilliant, and the prismatic hues more varied and gorgeous. The light appeared to cover the whole firmament, apparently like a luminous cloud, through which the stars of the larger magnitude indistinctly shone. The light was greater than that of the moon at its full, but had an indescribable softness and delicacy that seemed to envelop everything upon which it rested. Between 12 and 1 o'clock, when the display was at its full brilliancy, the quiet streets of the city resting under this strange light, presented a beautiful as well as singular appearance.[1]

POWER GRIDS COLLAPSE GLOBALLY

Today's power grid, only a few years old, has yet to experience a G5 solar storm on the magnitude of the 1859 event. And yet we are regularly reminded that solar storms can wreak sudden and devastating damage on the planet. Most of the time, the storms strike in one area, affecting only a few thousand or a few million people. So the event is "news" for a while, but shortly we all return to our busy lives and the next news story. We try not to imagine that any one of these solar strikes, perhaps the next one, could wipe out electronics on the entire planet.

This is denial, pure and simple, but also dangerous, because in the midst of it, we override harbinger after harbinger of the pain and suffering that *will* be visited upon us if our vital electronics are not hardened against a serious geomagnetic storm.

These storms do come often.

QUEBEC'S 1989 BLACKOUT

On March 9, 1989, astronomers at the Kitt Peak National Solar Observatory were looking up into the skies and first saw it coming. The sun spat a million-mile-wide blast of solar gas straight at the Earth. The scientists called it a "coronal mass ejection," and within minutes the Earth's atmosphere was swimming in powerful ultraviolet and X-ray radiation. The following day an even bigger—*thirty-six times bigger*—eruption could be seen through the telescope. It was on the dead center of the sun. And it hurtled out at 2.5 million miles an hour . . .[2]

Four days later, observers in the northern climes were treated to a spectacular auroral display—with the skies painted in vivid shapes that looked like dragons in battle. Some wondered if a nuclear first strike might be in progress.

Yet at 2:44:16 a.m., all was well at Canada's Hydro-Quebec Power Authority. Engineers watched the loads come and go as they always do during off-peak hours. At that moment there was no sign that invisible forces were staging a pitched battle in the sky above, with vast cascades of charged particles and electrons flowing west to east, inducing powerful electrical currents. One second later, those currents found a vulnerability in Hydro-Quebec's power grid. As reported by Sten Odenwald on the Space Weather website:

> A 100-ton, static VAR capacitor Number 12 at the Chibougamau substation tripped and went off-line as harmonic currents induced by the electrojet flowing overhead, caused protective relays to sense overload conditions. The loss of voltage regulation at Chibougamau caused power swings and a reduction of power generation in the 735,000-volt La Grande transmission network. At 2:44:19 AM, a second capacitor followed suit at the same station. 150 kilometers away at the Albanel and Nemiskau stations, four more capacitors went off-line at 2:44:46. The last to fall at 2:45:16 AM was a static VAR capacitor at the Laverendrye

complex to the south of Chibougamau. The fate of the network had been sealed in barely 59 seconds as the entire 9,460-megawatt output from Hydro-Quebec's La Grande Hydroelectric Complex found itself without proper regulation.[3]

Within seconds, six million people in Quebec were without electricity and would be for another nine hours on a cold, wintry night with outside temperature at 19 degrees. People were trapped in darkened office buildings and elevators; traffic lights didn't work; heating systems were snuffed out.

Isolated power failures continued for another day as Hydro-Quebec worked furiously to restart all the interconnected power lines and transformers across their network. Service to six New England states was interrupted while reserve sources of power were brought online.

The situation in control rooms across the Northeast had been just as tense. They had come very close to going down, as well. Allegheny Power lost ten of its twenty-four capacitors and acted quickly to take them offline to avoid damage. Had they not, Maryland, Virginia, and Pennsylvania would also have been plunged into darkness.

In the end, the only things that prevented a continuing cascade of electrical shutdowns and a projected $6 billion catastrophe extending across most East Coast cities were those dozen or so heroic capacitors on the Allegheny Power network.

Not since the great Northeast blackout of 1965 had U.S. citizens been so dramatically reminded of our vulnerability.

And not just in the Northeastern states. As far away as California, garage doors were seen opening and closing without apparent reason. Factories making microchips had to close shop because of the ionosphere's magnetic activity. Satellites tumbled out of geostationary orbit for several hours—their gauges no longer reading true north.

EXPLODING PIPELINE IN THE URALS

Just three months after the Quebec incident, disaster struck the Russian Urals. A natural gas pipeline sprung a leak, detonated, and killed five hundred people. Pipelines corrode; that is known. And the electromagnetic fury of a solar storm is known to hasten the corrosive process. Was the Quebec solar flare of three months prior to blame? That cannot be determined, or, at least, hasn't been determined. But it could surely be expected to happen again. The Russians aren't the only ones *not* taking good care of their vital infrastructure.

AIR FORCE ONE . . . DOWN?

While President Reagan was en route to China, *Air Force One* suddenly, inexplicably, lost communications. With the Cold War raging, the Pentagon scrambled to high alert—fearing the worst. The outage was brief, but only later did we learn that a solar flare had been responsible. Imagine if the Secret Service and military analysts had misread the situation and tried to assign blame?

CANADIAN SATELLITES BLANK OUT IN 1994

In January 1994, two Canadian telecommunications satellites went dark during a major sunburst. Communications were disrupted nationwide. The first satellite was fixed in just a few hours and brought back online. The second satellite took six months to repair, at a cost of approximately $70 million.

The first satellite was Telesat's Anik E1. It was disabled for seven hours after some space weather slammed into it and created too much static electricity for the control electronics to work. Anik E1 provided communications for all of Canada; the Canadian press was unable to deliver news to 100 newspaper printing depots and 450 radio stations.

An hour after the Anik E1 recovered, Telesat's Anik E2 blanked out. Some 1,600 remote communities lost their TV signals as well as data services. The satellite's backup systems were also damaged, rendering the $290 million satellite useless. Six months and $50 million in repairs later, the satellite was restored to full operation.

In the postmortem incident review, investigators, company officials, and the Canadian government had to be frustrated to learn that the Anik E1 and E2 failures *weren't* caused by a geomagnetic storm maximum. They occurred during what the National Academy of Sciences called "periodic enhancements of the magnetospheric energetic electron environment associated with high-speed solar wind streams emanating from coronal holes."[4] In other words, just a little duster. No more powerful than a Jacuzzi swirly in comparison to the tsunami that will hit—sometime soon.

UNITED AIRLINES FLEET THREATENED IN 2005

In January 2005, another solar storm hurtled toward Earth. Air traffic controllers at United Airlines feared that their craft in the storm path would lose radio contact. So they diverted twenty-six planes to nonpolar routes—as a risk mitigation step. Diverting the aircraft increased flight time considerably, forcing pilots to make some extra, unscheduled landings, and costing the airline $2.6 million in added fuel costs, along with the passengers' frustration at missing connections. But it all ended well that time.

POTENTIAL TO TAKE OUT HUNDREDS OF SATELLITES

Satellites carry highly advanced electronic components that are surprisingly defenseless. When a satellite travels through a geomagnetic storm, the charged particles strike the spacecraft, causing areas of the bird to be differentially charged, leading to damage and sometimes complete failure of the satellite's electronic systems.

Disruption to satellites is no small thing, especially since the satellites that service commercial communications *also* service the military. There are currently about 600 commercial satellites and 270 military satellites crowded into orbital parking slots above the Earth. The commercial satellites alone are valued at $75 billion and produce more than $25 billion a year in revenue. They've become "mission critical" in our information age, providing news, education, entertainment, global cell calls, satellite-to-home television and radio, and distance learning to populations in remote areas.

Satellites are intimately involved in every aspect of commerce. They take information from cash registers in thousands of stores, send those data to regional distribution centers, and provide automatic inventory control and pricing feedback for retailers. They also make it possible for us to pay with credit cards in those stores, and to enjoy the convenience of rapid credit approval and processing.

Automobile manufacturers use satellite-based networks to update their dealerships and provide service crews with the latest information on auto service and repair procedures.

Satellites also act as a critical backup to land cable systems, which are necessary to restore services during emergencies, such as earthquakes or hurricanes, when land-based communications systems have been knocked out.

Satellites provide valuable information on meteorology and weather conditions—having a direct and vital impact on agriculture, oceanography, forestry, geology, and just about every environmental science project going in the country.

Military satellites are used for image mapping, navigation, intelligence, telecom, and distant early-warning systems. They are also used for verifying compliance with arms control treaties and to support military operations. Without these satellites in good working order, the Pentagon would be practically helpless in every forward combat position. The military does maintain backup communications capability. But so much is now reliant on satellite-provided uplinks that our nation's fighting forces would be seriously compromised if a solar flare fried our satellites.

SAVE THE SATELLITES!

The Pentagon is quite aware of our liabilities should satellites fail in any great number, and the brass have taken a number of precautionary measures. Unlike the foot-dragging we've seen from them on EMP and RF weapons, the military is on top of these satellite concerns. No doubt this is partly because the folks at the Pentagon tend to be realists. They know that advances in space monitoring technology now give us warning—if only in hours—but warnings nonetheless of an explosion on the sun and an approaching solar storm.

The Pentagon is currently upgrading the sensors on crucial military satellites. In coordination with the National Science Foundation and international partners, the DoD (Department of Defense) is throwing money and manpower at protecting its vital satellites. As well, the DoD is becoming expert at monitoring electron levels in the upper atmosphere—so it is ready to take action (such as taking a satellite offline in a growing storm) to prevent costly damage.

SEARCH-AND-RESCUE AND EARLY-WARNING SYSTEMS . . . GONE

Today's search-and-rescue and early-warning systems make good lifesaving use of over-the-horizon radar—bouncing signals off the ionosphere to monitor the launch of aircraft and missiles from long distances. A geomagnetic storm would degrade these systems, which could be disastrous, particularly in a war zone.

Today's navigation systems use low-frequency signals, which require knowing the altitude of the ionosphere's lower boundary. Aircraft and ships use these very low frequencies to determine their positions. But during a solar event, the altitude of the ionosphere's lower boundary can change rapidly—introducing errors as significant as several miles.

If alerted in time to a geomagnetic storm, navigators can switch to alternative nav systems. Space weather forecasting thus becomes very important in assessing the solar anomalies that might affect satellites and their functioning.

WARNING AGAINST AN ANGRY SUN

Scientists at the National Oceanic and Atmospheric Administration (NOAA) are now using an advanced solar storm telescope called the Solar X-Ray Imager (SXI) to generate real-time images of the sun's explosive surface. SXI takes a full-disk image of the sun's atmosphere once a minute. These timely warnings can give NASA, satellite operators, and critical infrastructure managers time to respond and secure their assets from weather disturbances on the sun. According to Ernest Hilder of the NOAA Space Environment Center, "The SXI will detect and provide positions for 70 percent more solar flares than current ground observations . . . By knowing flare longitude, a forecast can be made that would be accurate for a window of about 12 hours. Without the solar longitude of a flare, the time of maximum particle radiation cannot be accurately predicted and can vary over a range of 100 hours."[5]

These advances in solar storm forecasting may represent the government's best response to the electromagnetic threat under which we are living.

OUR MODERN WORLD WITHOUT GPS

Global positioning systems, or GPSes, are integral to aircraft, automobiles, mobile phones, and so many other electronic devices. An entire generation has grown only vaguely familiar with those paper maps—considering them curious artifacts of a time gone by.

The GPS has become an essential piece of modern commerce. GPS-based businesses were valued at only $13 billion in 2003 but are expected to top $1 trillion in value by 2017. Any short-circuiting of this industry would have catastrophic consequences for the global economy.

Yet GPSes are also affected by geomagnetic storms, which jumble radio waves from satellites to the ground, ships, aircraft, or other satellites. During an October 2003 geomagnetic storm, GPS stations lost tracking capabilities on some or all channels. Both U.S. and international systems went haywire. It will happen again.

ENTERING SOLAR CYCLE 24

The 1989 Quebec blackout and more recent solar storms have combined to paint a vivid picture of what can happen. We know we are fortunate that truly powerful storms have been rare, thus far. It takes quite a solar wallop to cause a Quebec-style blackout. But in fact we have been getting walloped about every ten years or so.

We have endured five "solar storm maximum" cycles since the end of World War II, and we are now moving from a minimum to a maximum environment. Explains Tom Bogdan, director of the Space Weather Prediction Center in Boulder, "The sun has an activity cycle, much like hurricane season . . . It's been hibernating for four or five years, not doing much of anything."[6]

The year 2008 earned NASA's "Sun's blankest year of the Space Age" label. It hadn't been as quiet since 1913, when there were no solar flares. In 2009, the sunspot count dropped even lower. "We're experiencing a very deep solar minimum," said solar physicist Dean Pesnell of the Goddard Space Flight Center.[7] Sunspots had all but vanished, and solar flares were virtually nonexistent. In effect, the sun was quiet.

Until now.

Scientists began recording solar sunspot activity in 1755; we have now entered the twenty-fourth cycle, or Solar Cycle 24—when the sun is at its angriest and capable of critically wounding all the planets that surround it.

Early in this latest cycle, true to form, the sun began firing plasma bursts. There were four major bursts in 2011. Any one of them could have been G5 storms—but we got lucky. As it happened, the geomagnetic storms were only G1s, the most minor. How this occurs is rather fascinating, and strikingly random.

To understand it better, think about how a compass works. The needle of a compass is a freely rotating magnet that aligns with the Earth's magnetic north pole. When a solar storm hits and the storm's own magnetic field happens to be pointing northward, the two magnetic fields repel each other. In this best-case scenario, the solar storm is sent streaming around the planet, and the geomagnetic storm is greatly weakened. In a worst case, a solar storm's magnetic field points southward and locks onto the Earth's magnetic field, shooting charged particles into the atmosphere and strengthening the geomagnetic storm.

Both NASA and the National Academy of Sciences say the biggest storm yet could occur between 2012 and 2014. This cycle could reach its zenith in 2013 and slam Earth with energy equivalents comparable to that released by the atomic bomb at Hiroshima.

We almost saw that very energy burst in January 2012 when a coronal mass ejection was unleashed toward Earth, just missing us. We essentially dodged our solar system's biggest bullet. According to Antti Pulkkinen of the Space Weather Laboratory at NASA, that plasma blast was "more of a glancing blow . . . Our simulations show potential to pack a good punch to Earth's near-space environment."[8]

NEW DISCOVERIES HEIGHTEN RISK AND IMPACT

Scientists at NASA have only recently discovered that Earth is even more vulnerable to solar flares than previously thought. They have identified a thick layer of solar particles inside Earth's magnetic field that could *amplify* the geomagnetic storm. New data from NASA's THEMIS satellite revealed a four-thousand-mile-thick layer of solar particles gathering on the outermost part of the magnetosphere—the protective bubble created by Earth's magnetic field. With this buildup, the magnetosphere could be compromised and unable to perform its traditional function to full capacity.

When the Earth was in a solar storm minimum, this was not an issue. But now, with the increased activity of a solar storm maximum, the reaction in the magnetosphere could be dramatic. How dramatic? That's the question occupying the world's top scientists.

The largest such group of scientists, from seventy-five universities nationwide, has joined with the National Center for Atmospheric Research (NCAR) in studying the potential for global mayhem. The august team's analysis is less than encouraging. According to David Sibeck of NASA's Goddard Space Flight Center in Maryland, "The sequence we're expecting . . . is just right to put particles in and energize them to create the biggest geomagnetic storms, the brightest auroras, the biggest disturbances in Earth's radiation belts . . . So, if all of this is true, . . . we're in for a tough time in the next 11 years."[9]

Worse, say scientists, "the next sunspot cycle will be between 30 percent to 50 percent more intense than the last one."[10]

If these scientists are correct, then the solar storm that could strike any month now could produce solar bursts not seen since February 11, 1958. On that day, the nation suffered a radio blackout, cutting us off from the rest of the world. And in Europe, widespread fires broke out. Coming at the height of the Cold War, many wondered if World War III had begun.

Back in 1958, we didn't have the sophisticated means of measuring solar activity that exist today. However, people knew *something* was happening when the northern lights were sighted three times . . . in Mexico.

Today, such a solar maximum would impact all the electronics that are so much a part of everyday life.

Mausumi Dikpati, leader of the forecast team at NCAR's High Altitude Observatory, has also made an assessment for the next eleven years, based on what she says is a conveyor belt on the sun.[11] It is similar to the ocean's conveyor belt on Earth, except the sun's conveyor belt is a current of electrically conducting gas that flows from the sun's equator to the poles and back again. Just as the great ocean conveyor belt determines Earth's weather, the solar conveyor belt affects weather on the sun by controlling the sunspot cycle.

Digging deeper into the sun's composition, we learn that "sunspots" are actually tangled knots of magnetism generated by the sun's inner dynamo. A typical sunspot lasts only a few weeks. Then it decays, leaving behind a "corpse" of weak magnetic fields. The top of the conveyor belt skims the sun's surface, sweeping up the magnetic fields of old, dead sunspots. These corpses are dragged down at the poles to a depth of 120,000 miles below the surface, where the sun's magnetic dynamo can amplify them. Once the corpses are reincarnated, they float back to the surface and become new sunspots. This conveyor belt effect takes some forty years to complete one loop. And when the belt accelerates its pace, a greater number of magnetic fields are swept up, and a future sunspot cycle will be more intense.

That is where we are now. University of New Hampshire scientist Jimmy Raeder sees a telltale "perfect storm" coming: "For reasons not fully understood, CMEs [coronal mass ejections] in even-numbered solar cycles (like cycle 24) tend to hit Earth with a leading edge that is magnetized north. Such a CME should open a breach and load the magnetosphere with plasma just before the storm gets underway. It's the perfect sequence for a really big event."[12]

By "really big" event, Dr. Raeder is speaking of a coronal mass ejection—a slow-moving gathering of billions of tons of charged particles slamming into the Earth's magnetic field to produce trillions of watts of energy.

It will fry everything electronic.

5

THE ROUTE BACK TO THE
NINETEENTH CENTURY

NOBODY REALLY KNOWS how our world will change . . .

We have only guesses, educated guesses, and fiction writers' fantasies to guide us, to prepare us, for an event that is not only inevitable, but is, in all likelihood, soon. So let's take a look, as best as we can, at the road on which we could find ourselves, taking us back to a way of life from an era for which few of us are prepared.

THE COUNTRY WE KNEW IS NO MORE

Though we seldom think about it, we live in an ever more dangerous world, with a fragile complex of systems holding it all together. And when this exceedingly complex organism we call civilization has the "civil" yanked out and replaced by "real," we all will come to a sudden *real*ization that we are woefully unprepared to live any other way.

Within an instant, we will have no idea what's happening all around us, because we will have no news. There will be no radio, no TV, no cell signal. No newspaper delivered.

Products won't flow into the nearby Walmart. The big trucks will be stuck on the interstates. Gas stations won't be able to pump the fuel they do have. Some police officers and firefighters will show up for work, but most will stay home to protect their own families. Power lines will get knocked down in windstorms, but nobody will care. They'll all be fried anyway. Crops will wither in the fields until scavenged—since the big picking machines will all be idled and there will be no way to get the crop to market anyway.

Nothing that's been invented in the last fifty years—based on computer chips, microelectronics, or digital technology—will work. And it will get worse.

The average family has only a week's worth of food in the pantry. What will they do when the food is gone and there are no prospects of resupply? Millions of Americans will be forced to go foraging for food. The first targets will be the stores, restaurants, and food-distribution warehouses. Foraging will become looting by the end of the first day—all across America—because desperation turns us all savage. Twenty kids on an island are *Lord of the Flies*; two hundred million hungry adults are a plague of locusts. Any home in the neighborhood thought to have food will be set upon.

Next, people will move on to farms that are close by. Looters will form gangs, arm themselves, and plunge deeper into farm country—running their vehicles on siphoned gasoline.

In just a few days or weeks, there will be outbreaks of disease and no ready antibiotics. The outbreaks will not take long to become epidemics and begin a slow and painful attrition that will take out all who have not prepared.

This is what we can expect, but so much will be uncertain. Any reader of speculative science fiction will find a hundred different interpretations of what could happen, what will happen, what may or may not happen, etc. But again, we can't possibly know until it happens. All we can do to properly understand what could become the bleakest future our nation has ever faced is tease out some facts from the larger body of fiction that accompanies tales of EMP strikes.

WHICH VEHICLES WORK?

There is a huge library of myth that exists around which gasoline-powered vehicles will or won't survive an EMP/RF attack, or geomagnetic storm. The most popular scenarios, often presented in movies or in books, is that cars or trucks with electronic ignitions made after 1980, or vehicles with computer-directed systems, will be turned into lumps of steel, glass, chrome, and rubber the moment an EMP strikes.

The fact of the matter is that no one really knows.

There have been tests—government sponsored and mediated—of a very, very small sample of vehicles, on the order of one out of ten million. Some of the tests disabled all of the vehicles; some of the tests disabled none. For the most part, the tests were considered inconclusive because they could not subject a large, statistically significant sample of vehicles to a full-on EMP or sun-plasma simulated shock wave. There are simply no testing facilities capable of running such tests. And few, if any, manufacturers would be willing to submit their brand-name vehicles to such testing—they don't want to sustain the damage that could be inflicted on their brand image.

For example, "Mercedes E Class is EMP Dud" would do little for the Mercedes brand, obviously!

Even the EMP Commission admitted that its testing was limited. Under controlled circumstances, it aimed only modest EMP pulses at autos—and continued doing so until some component of the vehicle failed.[1] That's not an unreasonable approach since, in most EMP scenarios, not every car or truck will take a direct hit.

Millions of vehicles will be garaged below ground or deep within office buildings; millions more will be sitting at home, engines off, systems down; tens of thousands will be in car dealerships or gas station garages for maintenance; thousands will be on car carriers, being delivered to showrooms; and many more will probably emerge unscathed for reasons that cannot be explained.

Will some vehicles quit right away? Yes.

When an RF gun or e-bomb is directed toward a stream of traffic on a morning rush-hour bridge, the proximity of the pulse will be fatal to a number of vehicles. On-board computer controls for fuel management will be jolted. Radios will be cooked. GPSs will be toasted to a crisp. However, despite what's in the popular fiction or sensationalized media accounts, the actual electronic ignition systems in many vehicles may not even blink when smacked by a multi-gigawatt pulse. That's because the ignition system in every car—from forty-year-old classics to sports cars to soccer-mom minivans—works because of sparks, big ones that supply the juice to fire the spark plugs that ignite the gasoline in the cylinders. Those well-timed sparks have to be contained in layers of wiring and insulating materials if they are going to deliver their lightning-like energy to the engine. Internal combustion engines are already shielded against EMP—though not for that reason; no one in the auto industry thought that far ahead. It is simply because an unshielded ignition system is inefficient, irritating, and downright dangerous to the driver. You don't want to have an unshielded spark under the hood of a car as gasoline flows nearby.

What is going to bring traffic to a halt is the failure of other engine-support systems that control fuel flow, exhaust recycling, and the like. And even at that, cars may only suffer in performance. "Check Engine" lights, "Tire Pressure Low," warnings, and other notices may come on, and then die as their circuits take a hit. Many of the electronic options loaded on today's cars will malfunction in the wake of an EMP discharge.

However, it takes only a few vehicles stalled on a bridge, in a tunnel, across railroad tracks, on an overpass, or in the middle of a major intersection to initiate a ripple of misery across an entire commuting grid.

And think of all the traffic-related electronics that are vulnerable to EMP: traffic lights; information signs; traffic management operations; emergency responders; tow-truck companies; and the communications system that won't be able to distribute information about clogged highways.

As traffic grinds to a halt, a few dozen older cars and trucks will continue on—albeit through a gauntlet of stalled vehicles—as if nothing happened. That's because there will be no effects on older cars that operate without electronics.

FUEL WILL BE IN SHORT SUPPLY

Eventually, most of the vehicles disabled by an EMP/RF attack will be fixable. New battery cables and electronic systems will restore most vehicles to full operation. The real damage will come from the swift unavailability of fuel.

Gasoline and diesel fuel will be in short supply within days. It's not that the tankers delivering fuel won't get to the stations; some trucks will keep running or be easily repaired after an EMP strike. It's the sources of the gasoline, if hit hard by EMP, that will find themselves unable to run the pumps that move the gas from the main storage tanks to the tank trucks. Recall the SCADAs in the previous chapters—those simple control devices that act like switches and monitors for millions of pieces of electronic equipment all around the globe. Unprotected SCADAs, improperly shielded or hardened from high-voltage spikes, will cease to work. And that means that the devices that open and close the valves controlling the flow of gasoline into and out of the storage tanks at refineries, tank farms, and underground tanks at your local gas station will simply not work. And with their demise, the supply of fuel to a needy nation will slow to a trickle, then dry up altogether.

If there is a bright light on this situation, it's that a solar storm will probably have very little effect on vehicles. The wavelengths of the pulses emitted by solar flares are too long to disrupt automobile engines. However, a severe solar storm can still bring down the power grid, and as the power grid goes, so goes the ability to pump fuel at a gas station. Your car is going to need gas eventually. So it is always a good idea to keep your tank as full as possible so you have some mobility while the emergency crews do what they can to get the juice running again.

DRIVERS WITH PACEMAKERS—A PROBLEM?

The pacemaker has been a lifesaving invention to three million people worldwide who have had them implanted. So what will happen to the pacemaker, intracranial hearing aid, and other internally placed electronics?

While the actual pulse of electrons or microwaves has little or no effect on the human body, electronic circuitry installed inside the body does not fare well when subjected to high-energy microwaves. The devices could heat up very quickly, turning surrounding tissue and blood into steam, searing the body at the site of the device, and doing irreparable damage to the user.

In a traffic situation, a pacemaker-wearing driver's death would only add to the confusion swirling around the sea of dead or dying machines.

GETTING THE NEWS OUT—OR NOT

News of the growing numbers of traffic jams, the long lines of people trudging home, the unrelieved waits for electronically crippled services at stores, hospitals, pharmacies, or local police stations, will go unreported after an EMP blast. That is not to say reporters won't cover the events, or that radio and television stations won't want to broadcast them; it's just that they won't have the ability to get the word out on the airways because all of their production electronics—digital studio equipment, like cameras, microphones, and computers—will be useless, rendered mute by the physics of the pulse.

As with older cars being best suited to survive, broadcasters with less-current equipment will be in the best position to get back on the air. Older-technology AM and FM radio stations could be up and running from emergency sites within days or weeks of an EMP burst. Once the gear has been repaired, stations could be operating, as long as their backup generators run and their fuel supplies hold out. Once again, fuel is the key.

PHONES WILL WORK BETTER—AND WORSE—THAN COMMONLY THOUGHT

If getting the news out is going to be difficult, if not impossible, what are the average person's chances of staying in touch with family, friends, and employers?

Not all that good.

Telecommunications is the lifeblood of our modern society. An amazing 82 percent of the nation, or 250 million Americans, subscribe to cell phone plans, according to the CTIA wireless association.[2] We are now a nation talking, texting, viewing, and downloading for hours each day. But suddenly there would be no sharing of information, no sending for help, no reporting of crimes, no alerting neighbors of impending crises.

Cellular networks would be especially vulnerable to EMP, even more vulnerable than landline networks, because cellular network equipment is more susceptible to damage and has a more limited backup power capacity at cell sites than at landline sites.

In the aftermath of 9/11, the civilian telecommunications network experienced higher-than-normal call volumes. Cell call attempts rose to twelve times the normal rate in the region. This spike in volume meant that many callers were unsuccessful in reaching their loved ones. They experienced delayed dial tones or "all circuits busy" announcements. They were relegated to texting, which proved successful for the most part.

In an EMP event, cell traffic would likely increase to at least four times the normal rate and last for days, thereby degrading telecommunications services. This would also affect any priority routing that key government officials would expect to have. So the government would in all likelihood shut down nonessential and civilian access. After an event, when service crews went out to restore field systems, the cell base stations would be the slowest to be restored because each station and microwave tower would require a manual fix.

To guard against a loss of electric power, most telecommunication sites do use a mix of batteries, mobile generators, and fixed-location generators. This backup capability is usually good for seventy-two hours—that's the target most operators shoot for. To work longer, the backup systems depend on either the resumption of electrical power or fuel deliveries. So a power outage of less than a few days won't cause significant loss of telecom services, but after two days the phone lines will quickly go dead.

EXTREMELY DIFFICULT TO TEST "SOLUTIONS"

There is no satisfactory way to test "protective systems" to know how they will perform in an RF or EMP strike. Testing such a thing in downtown New York is clearly out of the question. Testing it out in the Mojave Desert isn't going to reveal much. And most of the devices that are being used—in the hopes that they do work—were first created in the early 1960s, when the nuclear testing of the time revealed that electromagnetic fallout could become a giant military obstacle.

These devices may work in some situations.

But in the years after nuclear testing was banned, interest in EMP solutions waned, and most of the engineers who had become experts have since moved on. And we don't know if EMP-focused protection will work as well for RF threats. One weapons expert, Carlo Kopp, warns that EMP-focused protection will not help against RF well at all and that there is only one way to be fairly certain that key systems survive:

> Reports from the US indicate that hardening measures attuned to the behavior of nuclear EMP bombs do not perform well when dealing with some conventional microwave electromagnetic device designs . . . It is significant that hardening of systems must be carried out at

a system level, as electromagnetic damage to any single element of a complex system could inhibit the function of the whole system . . .

Hardening new building equipment and systems will add a substantial cost burden. Older equipment and systems may be impossible to harden properly and may require complete replacement. In simple terms, hardening by design is significantly easier than attempting to harden existing equipment.[3]

Kopp is essentially saying that our best hope in moving forward is to design all future electronics-based systems with hardening in place.

FEDERAL GOVERNMENT "CLOSED FOR BUSINESS"

As much as the private sector, and even more, the federal government is dependent on electronic communications, departments and agencies would be seriously hindered in an EMP attack. This is a good thing, in the view of many. But it does mean that a number of government-based financial transactions—from Social Security payments to veterans' pensions and automatic deposits for active-duty military and federal government workers—would be affected.

Many critical offices in Washington are now equipped with walkie-talkies and would be able to communicate with first responders at the city and county level. The president would have at his command some semblance of control over the senior levels of his administration. FEMA would continue to operate, as would the Corps of Engineers and Coast Guard. The FAA would have its hands full for a few hours if its regional radars shut down, but it would not be a prolonged event. Airplanes would eventually land, or crash, and the airways would be empty after the last plane was down. Getting military aircraft up and running—both for airspace protection and missions of mercy—would not be a problem. Moving those aircraft to areas of greatest need—such as a blacked-out area—would take place due to the existence of backup navigation devices in the cockpits.

The federal government's emergency preparedness varies from department to department. However, the general plan is for cabinet leadership and their essential staff to move to secure locations outside Washington and activate an emergency command center. This center was used extensively during 9/11, continues to be staffed on a round-the-clock basis, and is activated regularly for emergency exercises.

The Veteran's Administration, as an example, has often publicized that it is a national model for electronic records management and security. Records of veteran patients are said to be secure and capable of secure backup at off-site locations. However, an EMP event could make the transmitting of those records highly problematic.

The VA has a nationwide footprint of 153 medical centers and 800-plus clinics along with mobile clinics and pharmacies. These facilities would all be called into duty in an EMP disaster. Department of Defense hospitals would also be available to care for a portion of the civilian population.

The IRS, Social Security, and other data-rich departments and agencies employ significant records backup systems to ensure a continuity of claims and other benefits, as well as taxation processes. If one data center were wiped out, it's safe to assume that a backup data center would be able to make good on the original records in most cases. But the thing that matters most to people—the check in the mail, or automatic deposit— would not happen at least for some delayed period of time if the banking computers went down.

Just as the Constitution and the Declaration of Independence are protected, so, too, are the national databases. The machines that run them are thought to be hardened, but it's uncertain how vulnerable they would be.

IF YOUR CASH ISN'T UNDER YOUR MATTRESS, YOU'RE SCREWED

Virtually all economic activity in the United States and other developed countries depends on a functioning financial services industry. At its core are information technology and networks without which the industry cannot operate. That is to say, when the power goes out, our money, for all intents and purposes, disappears and becomes unreachable until the lights go back on. And not just the small amounts of cash we have saved in our local bank, but our investments, our 401(k)s, and our children's college funds.

The automation of the financial services industry has spurred the growth of wealth, greatly increasing the amount of day-to-day transactions. In the 1970s, the New York Stock Exchange on an average day would see 10 million shares traded. Today it's 3 *billion* shares, involving a global complex of millions of players. This rapid increase in financial complexity has created a huge vulnerability. In the assessment by the EMP Commission:

> The increasing dependence of the United States on an electronic economy, so beneficial to the management and creation of wealth, also increases U.S. vulnerability to an electromagnetic pulse (EMP) attack . . .
>
> Modern financial services utilities have transformed the national economy from a paper system into an electronic system. Examples of some key industry utilities include FEDNET, Fedwire, ACH, Clearing House Interbank Payments System (CHIPS), the Society for Worldwide Interbank Financial Telecommunications (SWIFT), the National Association of Securities Dealers' Automated Quotation System (NASDAQ), the NYSE (New York Stock Exchange), the New York Mercantile Exchange (NYMEX), and the Depository Trust and Clearing Corporation (DTCC).[4]

Some of these acronyms are well-known; others are not.

FEDNET is the communications system that connects all 123 Federal Reserve banks nationwide and the financial services industry—transferring funds in real time, performing real-time sales and record keeping for government securities, and serving as a clearinghouse for electronic payments. CHIPS is an electronic system for interbank transfer and settlement and acts as a clearinghouse for foreign exchange. The SWIFT sends secure international messages between stock exchanges, banks, brokers, and other institutions. And the DTCC settles securities trades for participant banks and is the largest securities depository in the world.

All of these vital financial systems are totally dependent on the nation's electronic system. As well, all records of financial transactions are stored electronically. The financial infrastructure itself is a network of electronic machinery ranging from telephones, mainframe computers, and ATMs to vast data storage systems.

Security experts hired by the financial services industry have officially stated that the bulk of the current systems are capable of withstanding a wide range of threats, referring to the system as "highly robust." Yet a number of natural and man-made disasters have proven otherwise:[5]

- In 1988, for example, a fire in the Ameritech central office in Illinois disabled long-distance telecommunications for the Chicago Board of Trade and other major institutions.

- In August 1990, Wall Street was blacked out for almost a week due to an electrical fire in a Consolidated Edison office.

- In April 1992, underground flooding in Chicago affected telecommunications and powers, leading to serious outages for a period of time.

- In the February 1993 World Trade Center bombing, there were massive disruptions throughout Wall Street. Numerous problems with facilities, systems, procedures, and staffs were encountered as firms scurried to recover, and some securities firms' operations were shut down temporarily.

With 9/11, the threat we face had escalated, obviously, and the attention given to our vulnerabilities ramped up, as well. The Federal Reserve Board took action. It created emergency preparedness procedures to be administered by the National Communications System. The bureaucrats were instructed by the Fed to implement "same-day recovery" capabilities for a number of vital financial operations.[6] The status of these requests from the Fed is unknown.

So, today we have to assume that an EMP attack on the U.S. financial infrastructure could achieve the simultaneous destruction of all data backups and backup facilities in all locations. Wealth, recorded electronically in bank databases, could become inaccessible overnight. Credit, debit, and ATM cards could become useless. Even reversion to a cash economy might be difficult in the absence of electronic records that are the basis of cash withdrawals from banks. Most people keep their wealth in banks and have little cash on hand at home.

The alternative to a disrupted electronic economy may not be reversion to a nineteenth-century cash economy, but to an even earlier economy based on barter.

EMERGENCY SYSTEMS IN STATE OF EMERGENCY THEMSELVES

The nation's emergency services are essential for maintaining law and order and protecting property and public health and welfare. Citizens rely on a prompt response when local fire, police, rescue, and EMT services are needed. In turn, these local systems are backed up by capabilities at the state level, such as state police and the National Guard, with further backup provided by Homeland Security and its Federal Emergency Management Agency (FEMA); the Department of Justice (which includes the FBI, the Alcohol, Tobacco, Firearms and Explosive Agency, and the Drug Enforcement Agency); and the Centers for Disease Control and Prevention.

To give you some idea of the demand for emergency services, an estimated 200 million-plus emergency calls are made to 911 a year. More than 600,000 local law enforcement officers, a million firefighters, and more than 170,000 emergency medical paramedics respond to those calls across the nation.[7]

Since 9/11, governments at all levels have focused on improving emergency services and preparedness. But their focus has been almost universally on previous terrorist threats—like generals fighting the last war. Little if any attention has been paid to EMP attacks, and most every emergency service will be overwhelmed in the aftermath of an EMP attack.

Police officers would be called upon to assist rescue workers in removing people from immediate areas, direct automobiles, and control the traffic jams that will presumably occur following a chaotic event. In many cases, local police will be able to enforce at least a bare minimum of law and order. But that will not last, certainly not in urban areas. As we saw in the aftermath of Hurricane Katrina in 2005, order broke down quickly.

WATER GOES AWAY—FAST

Following an EMP attack, trucks and railway systems would be unable to deliver food and water in a consistent and timely fashion. The infrastructure needed to supply water, which is also dependent on electricity, is vital for everyday life, not only to meet the needs of the population, but also for agriculture and industry.

To support this demand, we have constructed an impressive national array of some 75,000 dams and reservoirs; thousands of miles of pipes, aqueducts, and water distribution and sewer lines; 170,000 water treatment plants; and nearly 20,000 wastewater treatment facilities. Treated water is delivered by high-lift pumps to a distribution system through pipelines pressurized up to eighty pounds per square inch to consumers. The pumps also help maintain water levels in storage reservoirs. SCADA systems are essential to managing the flow of water for drinking, farming, industry, and sanitation. These SCADAs enable centralized control and oversight of any system problems and failures and help identify breakdowns quickly in order to make vital repairs. So it was not surprising that the EMP Commission studied the water infrastructure and concluded: "An EMP attack that disrupts or collapses the power grid would disrupt or stop the operation of the SCADAs and electrical machinery in the water infrastructure."[8] And while some water systems have emergency power generators, they would be dependent on the availability of the necessary fuel to run them, inevitably diminishing performance. The generators could stop altogether, depending on fuel availability and usage. Large water treatment plants consume so much electricity that backup generators are impractical and instead draw their electricity from local power plants. If the electric power grid were taken out by an electromagnetic pulse event, the water infrastructure would collapse, "cutting off the water supply or poisoning the water supply with chemicals and pathogens from wastewater."[9]

Basically, if these systems go down, water cannot be purified or delivered, and sewage cannot be removed. Period.

FOOD GOES AWAY—FASTER

The food supply infrastructure in America relies heavily on electricity and water. Water for crops in the growing fields is supplied by aquifers and reservoirs through electric pumps, valves, and other machinery—all requiring electricity.

Egg and poultry farms are mostly automated—with feeding, watering, and cooling systems all run by electronics. The same is true of large-scale dairy operations—the milking of cattle will have to again be done the old-fashioned way, by hand.

Food processing facilities are also dependent on electricity to run the machines that clean, sort, package, and can the various agricultural products.

Prior to distribution, meat and produce are kept in large, refrigerated warehouses. Refrigerated trucks and trains take these perishable products to market across the country, making food distribution highly dependent on the transportation infrastructure.

It quickly becomes clear that there is an intricate system in place for bringing food from field to table. And it all relies on electricity. What's more, this distribution system has become extremely precision-oriented, thanks to computer modeling and just-in-time management. Modern supermarkets can get by with carrying only one to three days of supplies. They expect to be resupplied every few days by the warehouses. Electronic databases keep track of inventories and automatically send messages to suppliers to replace as needed. This reduces the need for large on-site inventories. It also means that supermarkets are highly vulnerable to an EMP attack.

Not only would the current supplies on the shelves be depleted within hours by panicked buyers, but those stores with refrigeration units would find their perishable foods spoiling within twenty-four hours of an EMP event.

In the cities, many people keep only a little food in the house. More often, they eat out. They would be most vulnerable. Families that stock up with weeks' worth of food will fare better, of course, but without refrigeration, most food is soon garbage. Within days, just about everyone will be out on the streets, looking for something to eat.

In addition, more than 33 million Americans live in what the Department of Agriculture calls "food-insecure households."[10] In the old days, we called this poverty. These Americans cannot afford to adequately feed their families now. In the aftermath of an EMP event, they would be the most vulnerable and the most desperate.

WHAT IS THE EXPECTED DEATH RATE?

In the blink of EMP's eye, 313 million Americans could be sent back to a much simpler time. The new normal would be a life turned upside down, and exceedingly hard on the majority of the great nation's people, who are woefully unprepared for a rural lifestyle as it existed more than 150 years ago. Some estimates suggest that only about 10 percent of our population, maybe 30 million people, could survive through a full year after a worst-case scenario.

The question is, would *you be among them?*

6

WHY HAVEN'T WE BEEN TOLD?

RF AND EMP WEAPON CAPABILITIES have been well known, if not fully appreciated, since at least 1997, when retired U.S. Army lieutenant general Robert L. Schweitzer testified before the Joint Economic Committee. His top concern, as he told the senators and representatives, was the low barrier to entry.

Any terrorist or malcontent, Schweitzer argued, could easily obtain the component parts to build an RF weapon. What's more, RF weapons are available to other countries for the asking, without the fear of scrutinizing investigators or even the inconvenience of licensing authorities.

When he testified, Schweitzer called for drawing up a list of those technologies needed to make RF weapons and placing them on the Pentagon's "Militarily Critical Technologies List." This is an essential first step in seeking to control the pace of development by both allies and potential enemies.

Schweitzer also noted that Russia was actively selling RF technologies to unknown buyers as far back as 1979: "Users of the new weapons can be criminals, individuals or organized gangs of narco or domestic terrorists—or a determined, organized, well-funded foreign adversary, either a group or nation who hates us."[1]

Despite firing this warning shot across Congress at a timely moment, little came of it. A few of Schweitzer's recommendations were implemented, but only on a limited basis. There was no coordination between the Departments of Defense, State, and Commerce. And the latter two departments did not add RF weapon-making components to their critical technologies list—so little was accomplished as a result of Schweitzer's plaintiff efforts.

And today, that critical technologies "list" remains only as a reference, no longer being actively updated. There are no export controls in place. Part of the reason is surely economic—commerce is generally good, especially when there's money to be made. Part of it is political— the list could become a sticking point with allies in conducting business in a competitive world. Both the economic and political arguments will surely pale in the face of the *human* argument when an RF device is easily, finally, deployed.

The task of controlling these new devices is problematic. No matter what we do, there is nothing to stop other nations from building and selling these weapons to the highest bidder. Government scientists at the Army's Aberdeen Proving Ground have said as much.[2] They reported that weapons are being developed by the United States, as well as Russia, Ukraine, the United Kingdom, China, Australia, France, Germany, Sweden, South Korea, Taiwan, and Israel, as well as Iran, North Korea, and Cuba.

What's more, the number of countries with RF technology is increasing. And several of these countries are known to be collaborating in their weapons development. The more collaboration, the more the communications channels open up, the likelier that terrorists obtain these weapons and deploy them.

A VERY PECULIAR MEETING IN MOSCOW

Congress first became concerned following a most unusual meeting in 1999. Congressmen Roscoe G. Bartlett (R-MD) and Curt Weldon (R-PA) were in Russia for a G-8 conference. While there, the two men also conferred with Vladimir Lukin, a onetime national security adviser to Mikhail Gorbachev. Lukin mentioned offhandedly that if Moscow really wanted to hurt the United States, with no fear of retaliation, Russia would launch an SLBM, or submarine-launched ballistic missile, detonate a nuclear weapon high above the country, and shut down its power grid and communications for six months or so.[3]

With this, the two Congressmen began enlisting support for measures to harden our nation's defenses against a potential attack—whether from Russia or from the budding terrorist movement of the time. But their efforts were promptly derided by the elite power structure. A *New York Times* story dismissed the threat with a quote from Phillip E. Coyle III, a former head of Pentagon arms testing at the time, who sneered that some kind of "EMP lobby" had been formed and populated with kooks: "[They] extrapolate calculations of extreme weapons effects as if they were a proven fact [and] puff up rogue nations and terrorists with the capabilities of giants."[4]

Whether Coyle was angling for a job at a left-wing think-tank or he truly believed that the terrorists of the world were incapable of anything so audacious, fortunately others were on the move. For that matter, critics of such events hadn't served in defense or the intelligence communities or never had access to classified data on electromagnetic pulse and its effects.

By 2004, Congress had formed a commission to study the EMP threat and to make recommendations.

The more cynical in Washington would say at the time that "commissions" are only formed when the politicians have a problem they don't know how to fix, or intend to ignore. And there's a rich vein of truth in that. But an optimist would say we have to start somewhere. And so it was that a distinguished group of scientists, academics, and politicians were appointed to the "Commission to Assess the Threat to the United States from Electromagnetic Pulse Attack."

THE FIRST EMP COMMISSION TACKLES THE ISSUE

The commission was given a substantial agenda and complete resources to study the potential threat faced by the United States, and make recommendations for action. Specifically, the EMP Commission's charter said it would:

- Assess the risk of a high-altitude EMP threat to the United States from terrorists that have or could acquire nuclear weapons and ballistic missiles;

- Look at all threats perceived possible for fifteen years in the future;

- Determine the vulnerability of the U.S. military and civilian infrastructure in terms of emergency preparedness;

- Determine how quickly the U.S. could repair and recover from damages on military and civilian systems from an EMP attack;

- Establish the feasibility and cost of hardening critical military and civilian systems against such an attack.[5]

The commissioners reviewed classified documents and scientific findings on electromagnetic energy, as well as the advances by other nations in developing EMP weapons and hardening against those weapons. In its first report to Congress in July 2004, the EMP Commission did not mince words:

> EMP is one of a small number of threats that can hold our society at risk of catastrophic consequences. EMP will cover the wide geographic region within line of sight to the nuclear weapon. It has the capability to produce significant damage to critical infrastructures and thus to the very fabric of U.S. society, as well as to the ability of the United States and Western nations to project influence and military power . . .
>
> The primary avenues for catastrophic damage to the Nation are through our electric power infrastructure and thence into our telecommunications, energy, and other infrastructures. These in turn can seriously impact other important aspects of our Nation's life, including the financial system; means of getting food, water, and medical care to the citizenry; trade; and production of goods and services. . . .
>
> The U.S. has developed more than most other nations as a modern society heavily dependent on electronics, telecommunications, energy, information networks, and . . . financial and transportation systems . . . This asymmetry is a source of substantial economic, industrial, and societal advantages, but it creates vulnerabilities and critical interdependencies that are potentially disastrous to the United States.[6]

Clearly the commissioners were concerned about the EMP threat, as a result of having studied the potential impacts. The good news, in their view, was that "correction is feasible and well within the Nation's means and resources to accomplish."[7]

The bottom line of the 2004 report was that the U.S. government should "harden" the U.S. critical infrastructure, at an estimated cost of $10–$20 billion over twenty years. The report also recommended taking strong action on the nation's missile defense systems to ensure the best protection against a nuclear attack.

Four years passed. The recommendations of the EMP Commission were all but gathering dust on the bookshelf. Some actions had been taken, but of little note. So a second commission was impaneled in hopes of re-energizing the government and military to action. Again the commissioners dug into the data, trying to make a stronger case that would stand up and attract notice at the highest levels. When their report was officially delivered to the Congress, the commission's chairman, Dr. William R. Graham, offered this supporting testimony:

> It is the consensus of the EMP Commission that the Nation need not be vulnerable to catastrophic consequences of an EMP attack . . . [Unfortunately,] our vulnerability is increasing daily as our use of and dependence on electronics continues to grow in both our civil and military sectors . . .
>
> Some critical electric power components are no longer manufactured in the United States, and their acquisition ordinarily requires up to a year of lead-time in routine circumstances. Damage to or loss of these components could leave significant parts of the electric power grid out of service for months to a year, or more. There is a point at which the shortage or exhaustion of sustaining backup systems, including emergency power supplies, standby fuel supplies, communications, and manpower resources, leads to a continuing degradation of critical infrastructures for a prolonged period, with highly adverse consequences to our population and forces.[8]

Dr. Graham also warned that it will not be just civilians at risk from an EMP event:

> Given our armed forces' reliance on critical national infrastructures, (e.g., electric power, telecommunications, food and water), a cascading failure of these infrastructures could seriously jeopardize our military's ability to execute its missions in support of our national security. Projection of military power from air bases and seaports requires electricity, fuel, food and water, and the coordination of military operations depends on tele-

communications and information systems, that are also indispensable to society as a whole. Within the U.S. these assets are in most cases obtained by the military from our critical national infrastructures.[9]

In his concluding remarks to Congress, Dr. Graham reiterated that the situation need not be as dire as it appears. The most debilitating consequences of an EMP attack could be sharply curtailed if there was a coordinated effort by the private and public sectors. Not only that, but the costs of such an effort could be manageable: "The cost for such improved security in the next 3 to 5 years is modest by any standard," he said, "and extremely so in relation to both the war on terror and the value of the national infrastructures threatened."[10]

However, despite the warnings of Dr. Graham's commission in 2008 and the previous commission in 2004, little has been accomplished. Despite being presented with respectable information from its own commission, Congress has yet to pass comprehensive legislation addressing EMP vulnerabilities. Numerous bills have been introduced, but none was passed out of committee. Here are the main ones:

H.R. 2195, "A Bill to Amend the Federal Power Act to Provide Additional Authorities to Adequately Protect the Critical Electric Infrastructure Against Cyber Attack, and for Other Purposes," was introduced in the House by Rep. Bennie G. Thompson in April 2009. The bill called for the EMP Commission to consult with Homeland Security to issue "such rules or orders as are necessary to protect critical electric infrastructure against vulnerabilities or threats." H.R. 2195 died in committee.[11]

H.R. 4842: Also known as the Homeland Security Science and Technology Act of 2010, this bill called for a new "Protection of Critical Electric and Electronic Infrastructures Commission" to continue the work of the EMP Commission. Although it was approved by the House, the Senate did not vote on H.R. 4842.[12]

H.R. 5026 was also known as the Grid Reliability and Infrastructure Defense (GRID) Act of 2010. It was sponsored by Rep. Edward Markey, passed the House in a voice vote, and was sent to the Senate. This act would force new industry standards for protecting critical infrastructure from cyber or EMP attacks. The bill was never put to a vote in the Senate.[13]

H.R. 6471: In the final days of the 111th Congress, Rep. Doug Lamborn sponsored this "Bill to Require the Director of National Intelligence to Submit a Report on the Foreign Development of Electromagnetic Pulse Weapons." Each country with an EMP weapons program was to be identified and its program assessed in detail. The focus would be on determining if EMP weapons could be integrated into other countries' nuclear weapons strategies. The bill never made it past the Intelligence Committee.[14]

H.R. 668: In February 2011, Rep. Trent Franks introduced this one, also called the "Secure High-Voltage Infrastructure for Electricity from Lethal Damage Act." This act would set up procedures for protecting the critical electrical grid, and also establish reliability standards for the grid to survive a geomagnetic storm or EMP.[15] As we write, this bill has been referred to the Committee on Energy and Commerce, as well as to the Committee on the Budget, where it sits. According to GovTrack.us, the bill has a 3 percent chance of being enacted.[16]

From this quick legislative survey, we can see that the EMP threat has languished on Capitol Hill. Or worse, as we'll see, it has devolved into a political football.

POLITICIANS TAKING SIDES . . . BLINDLY?

Like so many other realms, important and otherwise, the electromagnetic realm has become a political snarl. Instead of giving the matter a serious review and understanding, political sides have been staked out. Liberals, who generally favor slashing defense spending, have lumped EMP and RF into "things to oppose." So it was that the Institute for Policy Studies (IPS), a prominent "progressive" think-tank in Washington and a key driver of the Obama administration agenda, labeled the findings of the EMP Commission as nothing less than reactionary nonsense from radical right wingers "seeking to spread alarm about the purported threat of EMP attacks, which would involve the detonation of nuclear weapons in the upper atmosphere to generate a pulse that would knock out electronics-based infrastructure, [and who] have repeatedly used the findings of this commission to advocate increased funding for costly weapons programs such as missile defense and push alarmist notions that 'rogue states' like Iran and North Korea pose an existential threat to the United States."[17]

The left-leaning IPS was apparently so eager to oppose any EMP spending that they didn't bother to read the EMP Commission's findings. The commission had not advocated an increase in missile defense spending. Instead it had presented a cost-effective plan for hardening the national electric grid and other critical infrastructures.

On the right, several congressmen and former congressmen have taken strong advocacy positions. Former House Speaker Newt Gingrich is one. He spoke at the American Israel Public Affairs Committee annual conference in May 2009, saying, "Three small nuclear weapons at the right altitude would eliminate all electricity production in the United States. Which is why I have said publicly that I favor taking out Iranian and North Korean missiles on their sites."[18]

A few months later, Gingrich spoke at a conference hosted by EMPACT—a leading EMP advocacy group—and he ramped up the urgency of his message: "[An EMP attack] may be the greatest strategic threat we face, because without adequate preparation, its impact would be so horrifying that we would basically lose our civilization in a matter of seconds."[19]

Among the media covering Gingrich's remarks were the dependably liberal scribes for the *New York Times.* From their coverage, it was apparent they cared only about the politics of EMP, not the merits. A December 11, 2011, article by William J. Broad was titled "Among Gingrich's Passions, a Doomsday Vision." With a headline like that, you know the article was a critical hit piece. It spoke of Gingrich "ringing alarm bells" and stopped just short of advising the former speaker to tune up his tinfoil hat.[20]

Perhaps that journalist's name, William J. Broad, sounds familiar. He was the fellow we met back in chapter 3 who had written a 1962 article in *Science*, warning of the very real EMP threat.

Apparently the threat was real only until a conservative spoke about it.

In a similar vein, a year earlier, in March 2010, that same *New York Times* ran an article by President Obama's science adviser John P. Holdren and John Beddington, titled "Celestial Storm Warnings."[21] The intent of that article was actually quite admirable: to alert the public to the threat of EMP resulting from a great geomagnetic storm.

So when the EMP warning came from the Obama White House, it was a featured story in the "Gray Lady." But when a very similar EMP warning came from former Speaker Gingrich, it was cagily dismissed by liberal leaders.

SCIENTISTS TAKING SIDES . . . ALSO BLINDLY?

Political spin used to be confined to, well, politics. But today it seems to suffuse every conversation about every issue. The *Times* was not content in 2009 and 2010 to merely savage the "pro-EMP" crowd, calling EMP a ginned-up excuse for conservatives to spend money on missile defenses. The *Times* also printed the caveat "Many nuclear experts dismiss the threat."[22]

These experts—many in number, apparently—remain mostly unnamed to this day. But the *Times* quoted them—emphatically insisting (1) that America's current missile defense system would be able to thwart any attack, and (2) that countries like Iran and North Korea were not to be worried about because they are "at the kindergarten stage of developing nuclear arms."[23]

With some digging, these "experts" in the *Times* story could be identified. One was Yousaf M. Butt, an astrophysicist some, but not a national security analyst by any stretch of the imagination. Nonetheless, the good Dr. Butt had an opinion: "If terrorists want to do something serious, they'll use a weapon of mass destruction—not mass disruption. They don't want to depend on complicated secondary effects in which the physics is not very clear."[24]

Dr. Butt could barely contain his contempt for the EMP threat, which is probably why the *Times* plucked him out relative to issue an opinion on this important matter. And in the same year Dr. Butt was holding forth, three things were making news in less-partial publications:

1. Criminals were using RF to rob jewelry stores.
2. Satellites were being damaged by solar flares.
3. Terrorists were searching the Internet for plans to build RF weapons.

AND THE PENTAGON TAKING SIDES . . . BOTH SIDES

There is general agreement at the Pentagon that EMP and RF weapons could cause a lot of damage—even a catastrophic loss unprecedented in modern centuries. But the agreement stops there.

Senior military officials disagree heartily on the best way to respond. If there is any consensus among the ranks, it is this: Our nation's outstanding missile defenses are shield enough against any attack, including an electromagnetic one.

The Missile Defense Agency maintains an arsenal of ground-based interceptors at the ready, and capable of flying into space in just seconds to blow away enemy warheads. The agency believes that defeating an EMP attack would be as straightforward as any other defense of the continental United States. Agency spokesman Richard Lehner outlined the official position: "It doesn't matter if the target is Chicago or 100 miles over Nebraska. For the interceptor, it's the same thing."[25]

Dr. Lehner didn't outright dismiss the potential threat of an EMP attack, but he did categorize it as "pretty theoretical."[26] He pointed out that Iran's missile bases keep blowing up and North Korea's rockets can't seem to launch. He insisted that they would have to perfect big rockets and secret ways to enter U.S. airspace undetected. If they could succeed at that, our missile defenses would then spring into action.

This kind of talk is characteristic of Pentagon thinking—concerned only about the big-scale, Armageddon possibility—despite how all the attacks against the United States for more than fifty years have been smaller-scale, and often devastating at that.

Lehner's approach hurts America's strategic defense posture in another way—it turns every discussion into a conversation about missile defenses. Veteran defense policy analyst Frank Gaffney reminds us that a rogue nation seeking to tilt the battle to its advantage using EMP weapons does not even need long-range missiles:

> Short-range missiles launched off a ship would suffice to deliver a strategic EMP attack. Virtually all the world's bad guys—including not only North Korea and Iran, but the Islamist terrorist group Hezbollah as well—have Scud missiles that could perform this mission. What is more, Iran has already test-launched ballistic missiles off of ships and

launched another, the Shahab-3, in a manner that seemed to simulate a detonation at apogee. In other words, as soon as a nuclear weapon is available, Iran could be capable of waging an EMP attack.[27]

Gaffney pointed out further that our antimissile systems might not always be able to defend against missiles launched from the sea close to our shores: "The determining factor would primarily be the location and readiness of the Navy's missile-defense-equipped Aegis ships. Our west-coast–deployed, ground-based interceptors will be unable to do the job against short-range missiles fired off our east or Gulf coasts."[28]

MORE MILITARY MINDPOWER THROWN AT THE PROBLEM

In 1998, the CIA formed its own related commission—this one was charged with assessing the ballistic missile threat to the United States. Known as the Rumsfeld Commission, the bipartisan group included some of the most experienced defense experts in the nation, including Donald Rumsfeld, Dr. Paul Wolfowitz, Dr. William Schneider, James Woolsey, and Dr. William R. Graham.

This commission focused on missile development in Iran, Iraq, and North Korea, and the help they were receiving from Russia and China. They also looked at the full breadth of Western technologies that had been transferred in the Clinton years to China and Russia before being turned over to America's enemies. In one of the briefings, the commissioners revisited the 1998 Caspian Sea incident in which Iran launched a medium-range Scud missile from the deck of a cargo ship.

The briefer gave a theoretical example of a cargo vessel plying the Eastern Seaboard just like any of the hundreds of other commercial vessels on any given day:

> The cargo vessel could be out on the waters, going around our coasts through the normal channels used by other commercial cargo vessels. It could go back and forth for weeks or months and look routine as it would go in and out of various ports along the U.S. East Coast, ports in the Gulf of Mexico, or even visit ports on the West Coast. The ship would be registered with any number of countries to provide legitimacy and show a flag of a very neutral country, with no prior reference to any

other vessel that may have been used by countries of concern. The ship, however, actually could have a crew of highly trained missile experts who know how to launch SCUDs, especially from a ship.

The briefer concluded very simply: Any number of countries could be in this ship.

At the time, Iraq, Iran, and North Korea were thought to be years away from producing a nuclear weapon, even though they had nascent nuclear development programs. It was more likely, the briefer said, that the warhead would be either chemical or biological and exploded over a heavily populated area.

He said that the Scud would be suborbital, meaning it would not have to go into space but could just arc from the ship below what most radars off the United States were capable of picking up. Given that the United States didn't have any antimissile system at the time, the briefer said that the missile would not be intercepted and could be programmed to explode above a heavily populated city.

In effect, the briefer concluded, a rogue country could slip a covert ship-based missile launcher just off one of the U.S. coasts, launch its missile, and leave undetected.

Following this particular briefing, the members of the Rumsfeld Commission sat silent, looking at one another. Finally, former CIA director Woolsey spoke.

> If that missile was launched well off our East Coast at night, our radar system probably wouldn't be able to pick it up. The missile could be programmed to arc almost immediately after launch, to avoid detection. The ship then would continue on its way, and we would have a difficult time determining which ship was used to launch the missile. With chemical and biological weapons, accuracy would not be an issue. Just the fact that a missile with either a chemical or biological warhead exploding over one of the heavily populated areas in the megalopolis between Boston and Washington would have a serious psychological impact. Combined with the fallout from a biological or chemical attack, the effect of its detonation would be an unmitigated disaster.

VENEZUELA'S CHAVEZ IS PLAYING WITH FIRE, AND LOVING IT

Woolsey's prophetic outlook need no longer be confined to cargo ships plying the Eastern Seaboard. Iran announced in 2011 that its warships may begin using port facilities in Venezuela with President Hugo Chavez's blessing, thus maintaining a presence along the U.S. Gulf Coast. Iranian rear admiral Habibollah Sayyari said the warships were being sent near American soil in response to U.S. warships in the Persian Gulf, which he said was "illegitimate and makes no sense."[29]

The warships would be equipped with the latest missiles in Iran's arsenal. With vessels sitting in waters so close to the U.S. coast, the prospect is raised of a suborbital launch of a ballistic missile onto the U.S. mainland. However, the U.S. Navy would be shadowing the Iranian warships, ready to shoot down any such missile. So the chance of success on the Iranians' part would be slim. And they have to know that. So they have to know also that their odds of success would increase greatly if they used a covert commercial cargo vessel equipped with a missile launcher.

WORST FEARS—BOTH NORTH KOREA AND IRAN ON THE MOVE

Since the time of the Rumsfeld Commission's findings, their worst fears have been realized. Using Scud technology acquired years earlier from Egypt, the North Koreans then produced a single-stage, mobile liquid propellant medium-range ballistic missile, called the Rodong, with the help of the Chinese.

North Korea was long believed to lack the requisite technical skills to seriously threaten the United States. Now, they've proven that they can.

Meanwhile, Iran's nuclear ambitions have been long known, and feared. In November 2011, the UN's International Atomic Energy Agency (IAEA) released a shocking report on Iran's state of progress. It said Iran had conducted tests "relevant to the development of a nuclear explosive device."[30] Specifically, it charged Iran with working to miniaturize a Pakistani nuclear weapon and simulating nuclear explosions.

Iran, of course, denied everything. It called the U.N. watchdogs a "puppet" of the USA and said the findings were baseless and inauthentic.[31] Iran's envoy to the IAEA, Ali Asghar Soltanieh, said of the IAEA report:

"This report is unbalanced, unprofessional and prepared with political motivation and under political pressure by mostly the United States."[32]

Iran insists that its nuclear program exists solely for civilian and humanitarian purposes, such as running hospitals, despite the findings of the IAEA report:[33]

- Iran is developing fast-acting detonators that have "possible application in a nuclear explosive device, and . . . limited civilian and conventional military applications."

- Iran is testing the kind of detonators that are used for exploding a nuclear device.

- Iran is acquiring "nuclear weapons development information and documentation from a clandestine nuclear supply network."

- Iran is working on "the development of an indigenous design of a nuclear weapon including the testing of components."

The report stopped short of explicitly accusing Iran of developing a nuclear bomb. When the report was released, however, all eyes turned to Israel. Would it strike Iran's nuclear facilities, as it had struck Iraq's Osirak nuclear site back in June 1981, in hopes of slowing or even stopping Iran before it was too late?

As of the fall of 2012, there has been no strike and Iran continues its nuclear program. The Obama administration has said Washington will increase the pressure on Iran and slap it with additional sanctions if nuclear development continues there. However, Israel had declared that Iran was fast approaching a "zone of immunity" from a conventional military attack and felt that diplomacy had pretty much run its course, without halting the Iranian nuclear program. As for the UN Security Council, it has already voted four rounds of sanctions against Iran for refusing to halt uranium enrichment. Nevertheless, Iran believes that as a signatory to the Nuclear Non-Proliferation Treaty and as a member of the International Atomic Energy Agency, it has a "right" to develop nuclear energy and undertake the enrichment that goes along with it.

At this writing, Iran has enriched uranium up to 20 percent, which can be used for medical research. At the same time—and the thing that worries Israel, which isn't a signatory to the NPT but also is assessed to have nuclear weapons of its own—the amount of enrichment at low levels will make just that much more uranium available to enrich to weapons grade of more than 90 percent should Iran achieve that level. If this is true, then Iran would have enough uranium to make some eight nuclear weapons.

The curious thing about the NPT, however, is that it doesn't forbid a signatory from developing components for a nuclear weapon—that country just can't bring those components together to make a nuclear bomb.

There remains disagreement between U.S. and Israeli intelligence analysts as to whether Iran has achieved that level or indeed is working to do so, mindful of an earlier event in February 2003 when the United States made a major presentation showing "evidence" of weapons of mass destruction in Iraq.

A month later, the U.S. invaded Iraq to destroy what was thought to be operational WMD, but found none. Could history be about to repeat itself?

WENT TO SYRIA

WHERE WE ARE, AND WHERE WE NEED TO BE

In an enlightening paper, "Before the Lights Go Out: A Survey of EMP Preparedness Reveals Significant Shortfalls," Jay Carafano, Baker Spring, and Richard Weitz wrote:

> America—at all levels of governance—is unprepared for an EMP attack. Despite the clear recommendations of both the 2004 and 2008 EMP Commissions, U.S. government agencies have not taken planning for their response to an EMP attack out of the theoretical stages. This is especially alarming considering the official consensus on the severity of the threat and on appropriate solutions as articulated by the EMP Commission, the other aforementioned commissions, and the overwhelming majority of the expert community. DHS and DOE have both independently identified the United States' vulnerability to an EMP attack, but have neither created emergency management plans nor taken action to better protect critical U.S. infrastructure from attack. DOD has begun to adopt the recommendations of the 2004 EMP Commission, but U.S. forces still remain vulnerable. State and local governments remain unaware and unprepared for the threat of an EMP attack. Current priorities for the U.S. are:

- **BUILD COMPREHENSIVE MISSILE DEFENSES.** Maintaining the capacity to interdict nuclear-tipped missiles is the most effective measure to guard against a HEMP attack. The U.S. missile defenses are not keeping pace with the proliferation of threats. It is time to reverse course. Establishing a robust ballistic missile defense is the most effective means of addressing the future threats to the U.S. and its allies resulting from the proliferation of missile technology and weapons of mass destruction. The U.S. must pursue missile defense programs that can intercept missiles in the boost and ascent portions of flight. Among these programs are the Airborne Laser, which is a modified air-to-air interceptor missile, future versions of the Navy's Standard Missile-3 (SM-3) interceptor, and, above all, reviving the development and deployment of space-based interceptors.

- **DEVELOP A NATIONAL PLAN TO RESPOND TO SPACE WEATHER EMERGENCIES.** As a 2008 report by the National Academies, "Severe Space Weather Events—Understanding Societal and Economic Impacts," makes clear, "Modern society depends heavily on a variety of technologies that are susceptible to the extremes of space weather—severe disturbances . . . driven by the magnetic activity of the sun." The first step in addressing this issue must be educating the public and policy communities at the federal, state, and local levels about the risks and response options. Additionally, any effective plan will require enhanced, reliable long-range space weather forecasts.

- **FORGE A BIPARTISAN CONSENSUS IN CONGRESS TO ACT ON THIS ISSUE.** The response to the EMP Commission's findings has been uneven within the United States government, with the Department of Defense taking the initiative and the Department of Homeland Security apparently sitting idle. Congressional inaction has contributed to this uneven response.

- **ESTABLISH BILATERAL PARTNERSHIPS WITH OTHER NATIONS.** If the unthinkable happens, the U.S. and other developed nations must be able to accept foreign aid in the event of catastrophes. The U.S. should consider hosting international disaster exercises to increase the ability of countries friendly with the United States to readily accept aid from one another when disaster strikes. For some critical infrastructure the U.S. should promote establishing an industry-led, multi-national rapid-response capability. Such a capability should be able to respond worldwide. Further, this could provide an effective mechanism to share best practices and integrate responses. This capability should be funded and controlled by the private sector to respond to threats to shared international critical infrastructure, such as telecommunications and the Western Hemisphere electrical grid.

"The effects of large-scale HEMP have been studied over several years by the Defense Atomic Support Agency, the Defense Nuclear Agency and the Department Special Weapons Agency, and is currently being studied by the Defense Threat Reduction Agency (DTRA)," said Clay Wilson of the Congressional Research Service. "However, the application of the results of these studies has been uneven across military weapons and communications systems. Some analysts state that U.S. strategic military systems such as intercontinental ballistic missiles and long-range bombers may have strong protection against HEMP, while most U.S. weapons systems used for the battlefield do not, and that this uneven protection is undoubtedly known to our potential adversaries.[34]

From this summary analysis, we see that an EMP catastrophe would be devastating but also, accidents notwithstanding, avoidable. The means to address and mitigate the dangers to critical infrastructure are at hand. All the United States needs is a greater understanding of the danger—and the determination to act.

7

WHAT IS THE CURRENT THREAT ASSESSMENT?

W E'VE SEEN IN THESE PAGES that the U.S. government and major business interests have little desire to tackle the EMP and RF threats, and that among the current national leadership, there is almost an antipathy toward the entire subject. As if they're above it all. This is especially troubling because the threat is even broader than we've detailed thus far, and it is already touching each of our lives in very direct ways.

A number of people interested in obtaining and using RF weapons are not terrorists or even state sponsors of terrorism. They are corporate espionage agents, technology-savvy criminals, disgruntled employees, and disaffected or disturbed "lone wolves."

Indeed, the FBI has identified these "lone wolves" as serious threats to homeland security. A number of them have been snagged in undercover FBI sting operations, caught trying to purchase large-scale weapons and explosives for their wicked schemes. Some of these are societal dropouts; others are extremists or anarchists aiming to topple our government. But all of them have been presented with a new—and for them, advantageous—reality.

Until recently, when they wanted to obtain weapons and explosives, they often needed to use an intermediary. That intermediary has been, in many cases, a well-trained undercover agent. With RF weapons, however, there is no reason to reach out to an intermediary or dangerous middleman.

Anyone with some technical smarts can purchase the components and complete the assembly without involving a soul, making any undertaken attack a total surprise. A few examples:

- In the Netherlands, one such character used an RF weapon to disrupt a local bank's computer network. His reason? The bank turned him down on a loan. The weapon was briefcase-size, built from schematics found on the Internet. Banking officials had no idea what hit them, no understanding of how or why their expensive, high-level security system could be so easily breached.[1]

- In St. Petersburg, Russia, a jewelry store was robbed after an RF generator was used to disarm the store's alarm system. Authorities reported that the RF device was no more complicated to assemble than a home microwave oven.[2]

- Again in Russia, the Chechen rebel commander Salman Raduyev used RF to disable police radio communications during a raid.

- In Japan, two *yakuza* criminals used RF to falsely trigger a win on a pachinko machine.[3]

- In March 2001, there was a mass failure of keyless remote entry devices on thousands of cars and trucks in Bremerton, Washington. Nobody could figure out why. The FBI got involved. The FCC arrived on the scene. Our taxpayer dollars were hard at work trying to figure out the culprit. Turns out, the USS *Vinson* aircraft carrier had just pulled into harbor, and its onboard electronics seemingly fouled the remotes.[4]

Each of these incidents, though relatively small in scale, underscores the potential for mischief—especially knowing how easily the weapons can be obtained, and used.

ADVERSARIES UNDERSTAND JUST HOW VULNERABLE WE ARE

Nations that are hostile to America know something: They know that each new technological breakthrough we celebrate makes us even more vulnerable, and there will come a tipping point in their favor. Says EMP

Chairman William Graham, "Our vulnerability is increasing daily as our use of and dependence on electronics continues to grow in both our civil and military sectors. The impact of EMP is asymmetric in relation to the potential antagonists who are not as dependent on advanced electronic technologies."[5]

Our nation is arguably the most dependent of all nations on communications technology. We rely on electric power, electronics, telecommunications, information networks, and an extensive set of financial and transportation systems that leverage modern technology. But this dependency, if disrupted, could lead to devastating ends.

In this connection, a Third World adversary could see itself as the proverbial David finding that he has a tool with which he can defeat an otherwise impregnable Goliath.

In fact, a number of weapons authorities argue that the advent of easy-to-obtain and low-cost RF weapons puts the United States at greater risk of an EMP attack than during the height of the Cold War.[6] Then we faced an identifiable enemy; now we face an ambiguous motley of bad actors who are difficult to deter—since they can come from just about anywhere.

IRAN'S DOWNING OF A U.S. DRONE—TRUE OR NOT?

The most public and controversial use of advanced electronics warfare was very recent—December 4, 2011. On that day, senior military officials in Iran boasted that they had successfully downed an RQ-170 Sentinel drone, carrying some of our finest American technology. The drone had taken off from Afghanistan to spy on the Islamic republic. Iranian officials proudly displayed the drone, which had only minor damage.

By all appearances, the "drone down" incident caught the U.S. military by surprise, and a Pentagon spokesman insisted that the drone had experienced mechanical failure. At the same time, Iranian officials were insisting in highly staged TV news reports that their superior scientists had used new microwave technologies to "trick" the electronics on board the drone and caused it to land gently on Iranian soil.

Both Iran and the United States were motivated to conceal what they really knew about this incident; so who was telling the more accurate story?

We cannot be certain. We do know that high-ranking Iranian politician Esmaeil Kowsari could barely conceal his glee in his public taunting of the U.S. military. On TV reports viewed around the world, Kowsari said, "The plane was designed to fly on SAT [satellite] orders from an initiation base. If for any reason contact with the initiation base—in this case Afghanistan—was lost it should respond to orders given by another base in the USA. If contact with both of the bases is lost, then it should return immediately to the initial base. If it cannot, self-destruction is the ultimate choice. Despite this, the Iranian electronic war hackers were not only able to hack the SAT controlling the drone and sever contacts with the Afghanistan base, but they were also able to simulate their base as the USA base for the plane and as the higher ranking base forced the plane to land without destroying itself, the artificial intelligent-based software did not even suspect that it is being controlled by the Iranian forces."[7]

In the war of information, Mr. Kowsari then ratcheted up his condemnation of the United States and issued the most thinly veiled of threats: "Since the algorithm and logic behind other systems are much simpler, the Iranians may be able to control all radars, satellites, planes, ships, tanks, rockets, cruise missiles of the US and NATO throughout the world. They may be able to even control American soldiers, who are driven to fight using satellite controlled infrared systems."[8]

Mr. Kowsari could have been blustering. This could be empty bravado, a taunting intended more for his own political constituents than for the United States. But perhaps not. We have seen enough evidence now to suggest, if not confirm for certain, that Iran has obtained technologies that could completely alter the field of future battle. As threats to U.S. military supremacy go, it's hard to imagine a more destabilizing and demoralizing scenario. And if Mr. Kowsari was hoping to scare the Pentagon, he succeeded on some level.

The Pentagon has, in fact, known about this potential for harm for more than a decade. In 1999, Lieutenant Colonel John A. Brunderman published a forty-one-page paper titled "High Power Radio Frequency Weapons: A Potential Counter to U.S. Stealth and Cruise Missile Technology."[9] Nobody doubted Brunderman's credentials on the subject matter. He was at the time a senior analyst at the Center for Strategy and Technology of the Air War College. In his paper, made available on the Internet, Brunderman went into great detail about the potential for using high-power radio frequency in both offensive and defensive military campaigns.

Brunderman's paper was so thorough and instructional that any reader could come away with tactics for downing our stealth aircraft, including our drones. Where the paper did lack detail, it was sufficiently robust enough that talented engineers could extrapolate their way to the construction of systems capable of thwarting our finest radio frequency technologies.

SO MUCH FOR TECHNOLOGY SECURITY

This skirmish with Iran was obviously embarrassing to the Pentagon, but it also suggested a larger vulnerability. Brunderman had predicted this vulnerability as well: "Should an adversary ever devise the means to significantly increase our own casualty rates or even precipitate our killing of their civilians, intentionally or not, they have a good chance of altering the course of the war, especially in small scale conflicts where U.S. national survival is not at stake. High power radio frequency weapons are a new technology on the horizon that might provide just such a means."[10]

Brunderman could not have been more emphatic in his warning that our adversaries would obtain the capability to alter the course of the war using a little technological slingshot. And in the context of the downed drone, the altering could be profound and substantial.

Suppose the unmanned drone had been carrying a warhead and the Iranians had used RF to scramble the drone's GPS, causing it to target that warhead at a heavily populated civilian target. The United States could—and would—be blamed for the tragic deaths. Nation after nation would stand at the UN to decry the imperial hubris of the American superpower. We would be charged with war crimes. Meanwhile, the finest minds at the Pentagon would have no idea how the drone went rogue, or at least no way of proving malice.

MILITARY DOWNPLAYING RF FOR DECADES

The true lethality of RF devices was discovered by the Pentagon *by accident* in July 1967 on board the aircraft carrier USS *Forrestal*, engaged in combat operations in Vietnam at the time. The ship's radar had been improperly shielded, and a blast of energy was thrown off, slamming into an F-4 fighter and an ammo stockpile nearby. The resulting explosion and fire were the worst on any U.S. carrier since World War II, killing 134 sailors and injuring 161 others; damaging or destroying twenty-one aircraft; and costing $72 million in damages, equal to $474 million in today's dollars.

The Navy spent seven months repairing the *Forrestal* and upgrading with equipment to withstand future explosions and fires. But they spent very little time studying all the ramifications, both offensive and defensive, of radio-frequency technologies.

Decades later, as the Cold War wound down, President Clinton cut defense budgets drastically, and what little research and development were under way at the Pentagon became a low, even nonexistent priority. Precious little attention was paid to hardening the electronics in weapons systems. As more countries developed RF microwave technology, hardening efforts still didn't keep up. It wasn't until years later, long after it became clear that the Soviets were perfecting a new wave of military technologies, that the United States, once again in a Sputnik-style catch-up mode, began to pursue its own research and development.

RUSSIA'S 1999 EXPERIMENT

According to Lieutenant Colonel Brunderman, the Soviets constructed an experimental gigawatt-level microwave emission triode in 1979. Such a device could be expected to amplify the radio technology, and still keep it under control. By 1983, the Soviets had perfected a 100-megawatt millimeter wave band transmitting tube—which could for the first time produce a microwave beam powerful enough to be called a "weapon." Then the Soviet Union collapsed, and research efforts fragmented. But remnants of the work continued on in the former Soviet countries. For example, in one reported experiment, a microwave pulse was injected into the electrical power lines of a building, destroying all the computers therein.

At one point, the Russians moved beyond experimentation. They began "freely sharing their knowledge with those who can pay for it," said Brunderman.[11] The commercial version of their RF weapons, called the Ranets E High Powered Microwave, is available on the open market to anyone with a checkbook.[12] Therefore, it can be logically assumed that a number of countries that mean us harm have purchased these RF weapons.

And for those not wishing to do business with Russians, there are the Germans. The firm Diehl sells briefcase-sized RF weapons. More recently, as we've discussed, U.S. companies have begun selling these RF devices, as well. These open-market sales become more disconcerting when they are linked to activities of the underworld . . .

A.Q. KHAN'S NUCLEAR SMUGGLING NETWORK

In July 2008, EMP Commission chairman Graham testified before Congress on a range of threats, including the nuclear capabilities of North Korea. He detailed how UN investigators had found the blueprints for an advanced nuclear weapon in the possession of Swiss businessmen Marco Urs and Friedrich Tinner—both known to be affiliated with the smuggling network of Pakistani nuclear scientist Abdul Qadeer Khan, aka, A. Q. Khan.[13]

Two years earlier, Swiss police had acted on U.S. intelligence to seize computers believed to belong to Tinner family members. On those heavily encrypted computers were some 1,000 gigabytes of detail on building a miniature nuclear device that could be fitted on the type of ballistic missiles used by North Korea, Iran, and a dozen other developing countries.

At about the same time, a separate UN report had confirmed that the A. Q. Khan smuggling ring had succeeded on two fronts: They had sold nuclear bomb-related parts to North Korea, Iran, and Libya; and they had acquired blueprints to miniaturize an advanced nuclear weapon. The UN report suggested that those blueprints were also shared with a number of hostile countries and possibly with unidentified rogue groups.[14]

The UN report was authored by David Albright, a well-respected nuclear weapons expert with the Washington-based Institute for Science and International Security. Knowledgeable about the A. Q. Khan network, he wrote, "These advanced nuclear weapons designs may have long ago been sold off to some of the most treacherous regimes in the world . . . These would have been ideal for two of Khan's other major customers, Iran and North Korea . . . They both faced struggles in building a nuclear warhead small enough to fit atop their ballistic missiles, and these designs were for a warhead that would fit."[15]

Dr. Albright *did not say* that the plans for a mini-nuke had for certain been delivered into the hands of hostile nations. However, the previous testimony of Dr. Graham strongly corroborated Dr. Albright's suspicion that Iran and North Korea did acquire the plans. And as head of the EMP Commission, Dr. Graham had access to the highest-level intelligence sources. Dr. Graham was on highly credible footing when he told Congress, "Advanced nuclear weapon designs may already be in the possession of hostile states and of states that sponsor terrorism . . . It would be a mistake to judge the status and sophistication of rogue nuclear weapon programs, based solely on their indigenous national capabilities, since outside assistance may well have been provided."[16]

HOW PREPARED IS THE U.S.?

It's the most important question of all: *Is our nation's military prepared for an EMP attack?*

The verdict thus far in our investigation has been a resounding "No!" But let's look more closely at the specific programs the Pentagon has undertaken, to better understand what kind of national security we're getting for our taxpayer dollars.

In 2010, the Army ran tests on the air traffic control towers under its control, focusing on "high altitude electromagnetic pulse testing." Based on those tests, it sought 2012 funding for some protection projects; that is a sign that something is being done, though the Army has not yet released details of those "protection" projects.[17]

The Defense Logistics Agency (DLA) is studying improved technologies for shielding mission-critical 3-D electronics. It secured funding of $4.8 million for its Microelectronics Technology Development and Support Project in 2010. But given the DLA's $296 million overall budget, a few million for EMP testing does not shout "committed."[18]

The Air Force increased its budget to $50.5 million in 2011 to study what happens to electronic circuits when blasted by "electromagnetic waveforms."[19]

It fries them.

The Air Force wasn't content to just study the obvious; it also budgeted $22 million to actually do something. The money was earmarked to develop "novel materials for electromagnetic interactions with matter for electromagnetic pulse (EMP) high power microwave, and lightning strike protection" for aircraft, spacecraft, launch systems, and missiles.[20] By this, we think they mean "better jamming techniques."

To its credit, the Air Force has in fact moved to harden its B-52 and B-2 bombers, F-15s, and AWACS from EMP attack and geomagnetic disruption.[21]

THE GREAT CONTRADICTION—WE DON'T PLAY DEFENSE, ONLY OFFENSE

While the Pentagon has done little to protect our nation against EMP attacks, it is clearly aware of the offensive capabilities of EMP and has used them in combat in Iraq.

As well, plans are under way to field even more exotic electromagnetic weapons. In a recent *Economist* article, an unnamed Pentagon source spoke about new-generation EMP weapons that do not require a nuclear explosion, but instead use a type of radar called an *active electronically scanned array*, or AESA.[22]

An AESA, acting as normal radar, broadcasts microwaves over a wide area or a single point—depending on the need. A beam from an AESA scrambles its target's electronics, similar to the effects of an EMP from a nuclear detonation. Depending on the AESA's strength, it can be effective at considerable distances, without endangering the aircraft equipped with the AESA radar.

There are small and large AESAs. The larger ones fit on ships, while smaller ones fit in the nose of the F-22 Raptor and the latest F-35 stealth joint strike fighter. AESA's capabilities remain classified. But we know this weapon on board an F-35 can halt air-to-air and surface-to-air missiles. And ground- or ship-based AESAs can attack ballistic missiles and aircraft.

The Navy has added an offensive EMP capability to the Boeing Growler, a modified F-18 Super Hornet now known as the EA-18G airborne electronic attack aircraft. These EA-18Gs will be the cornerstone of the naval *airborne electronic attack*. They incorporate advanced airborne electronic attack avionics, and are capable of both suppressing enemy air defenses and initiating electromagnetic pulse attacks on enemy positions.

First used in Iraq in 2010, these jets are fitted with five pods, two

under each wing and one on the belly of the aircraft, containing electromagnetic weapons, surveillance, jamming, and intelligence-gathering
technology.

The Pentagon also contracted with Applied Physical Electronics, LC,
to develop what it calls the latest "compact directed energy solution." It is
an EMP weapon in a suitcase, and it "generates extremely high-amplitude
electric fields . . . suitable for affecting electronics and testing system vulnerabilities," according to the company's description.[23]

This is the extent of the Pentagon's known activities relating to EMP
and RF. In short, some very impressive advances on the offensive front,
but foot-dragging on defense. There are explanations for this.

It's true that Pentagon strategists are being forced to make budget
allocations in a time of cutbacks. President Obama has said that "after
more than a decade of war, it is time to focus on nation building here
at home."[24] So the Pentagon is operating in a gray zone. It has *many*
important projects to protect. And the EMP threat remains, in the view
of many, an existential threat. It is seen as almost too deadly to even
contemplate. A "game over" situation. That can be a scary proposition,
one that triggers the same kind of denial—though it shouldn't, not at the
Pentagon of all places—that it triggers in many everyday Americans. And
so, if there's a way for the Pentagon's decision makers to kick the EMP
can down the road and hope the bill comes due on someone else's watch,
then that's going to be the plan for many now working in the great maw
that is government.

THE ONLY REAL DEFENSE IS A MISSILE DEFENSE

It's clear that EMP and RF are difficult to protect against. We can harden,
and we should. We can wrap our most vital electronics in protective
sheaths, and we should. But if a hostile nation succeeds in exploding a
high-altitude nuclear missile overhead, our nation will be in great trouble.
That is why those who have studied this threat tend to agree on one thing:
the only real defense is a missile defense.

The best way to protect against a high-altitude EMP delivered by a
nuclear ballistic missile is our own missile defenses. Unfortunately, our
missile programs are being downgraded and steadily defunded. The

Obama administration made large-scale cuts to the missile defense program in 2010, and its proposed budgets for 2011 and 2012 will not make up for lost ground. The administration has also cancelled, or sharply curtailed, long-scheduled joint projects with U.S. allies, including the Airborne Laser program and missile defense systems in Poland and the Czech Republic. Lastly—and most egregiously, in the eyes of many defense analysts—President Obama has signed a new Strategic Arms Reduction Treaty with Russia that sharply restricts U.S. missile defense options.[25]

PENTAGON'S DOLLAR-AND-SENSE CONCERNS

In July 21, 2008, the Congressional Research Service issued its own report on the feasibility and costs of hardening our vital military systems. Written by technology specialist Dr. Clay Wilson, the report was titled "High Altitude Electromagnetic Pulse (HEMP) and High Power Microwave (HPM) Devices: Threat Assessments." Dr. Wilson concluded: "Hardening most military systems, and mass-produced commercial equipment including PCs and communications equipment, against HEMP or HPM reportedly would add from 3% to 10% to the total cost, if the hardening is engineered into the original design. To retro-fit existing military electrical equipment with hardening would add about 10% to the total cost."[26]

Dr. Wilson's report was not the first attempt to convince the Pentagon that the cost of hardening mission-critical systems would be small in comparison to the potential devastation. Back in 2004, when the *first* EMP Commission released its findings, one of the EMP commissioners observed that for at least forty years the DoD had resisted every serious discussion of EMP.[27] There was one simple reason for that: every time Pentagon analysts ran an EMP game scenario, it tended to end the game— the U.S. ended up in some nineteenth-century Hobbesian rendition of squalor and disease, with life nasty, brutish, and short.

Given how easily a terrorist can obtain an EMP weapon, this is truly frightening.

The problem is exacerbated by the current political environment and the Obama administration's efforts to shrink the U.S. military budget, with hundreds of billions in cuts slated over the next ten years. With shrinking budgets, it is at least understandable why the Pentagon would be reluctant to budget for a massive retrofitting of existing equipment to protect against an EMP attack. Yet there is no reason why the military cannot be designing future weapons systems with hardening technologies in mind. But this is not happening, as Dr. Ashton Carter, former under secretary of defense for acquisition, technology, and logistics, has pointed out: "The technical community should be exploiting tests and/or upgrades planned for operational hardware as vehicles to help rebuild and enhance the supporting technology base. The Task Force urged DTRA [the Defense Threat Reduction Agency] to engage in the planning for the March 2011 B-2 stealth bomber HEMP [high-altitude electromagnetic pulse] test to ensure that collected data supported code validation and development. Unfortunately that did not occur."[28]

It may appear that Dr. Carter is a lone sentinel, crying wolf. Not at all. His concerns have been echoed by a number of authoritative sources. One of these sources was the very first EMP Commission, which wrote back in 2004:

> EMP . . . test facilities have been mothballed or dismantled, and research concerning EMP phenomena, hardening design, testing, and maintenance has been substantially decreased. However, the emerging threat environment, characterized by a wide spectrum of actors that include near-peers, established nuclear powers, rogue nations, sub-national groups, and terrorist organizations that either now have access to nuclear weapons and ballistic missiles or may have such access over the next 15 years have combined to place the risk of EMP attack and adverse consequences on the U.S. to a level that is not acceptable . . . Our increasing dependence on advanced electronics systems results in the potential for an increased EMP vulnerability of our technologically advanced forces, and if unaddressed makes EMP employment by an adversary an attractive asymmetric option.[29]

Another authoritative warning has come from two former secretaries of defense, William J. Perry and James R. Schlesinger. In 2009, they headed up a bipartisan group of defense authorities to write "America's Strategic Posture: The Final Report of the Congressional Commission on the Strategic Posture of the United States." Commenting on the threat from EMP weapons, they wrote:

The United States should take steps to reduce the vulnerability of the nation and the military to attacks with weapons designed to produce electromagnetic pulse (EMP) effects. We make this recommendation although the Commission is divided over how imminent a threat this is. Some commissioners believe it to be a high priority threat, given foreign activities and terrorist intentions. Others see it as a serious potential threat, given the high level of vulnerability. Those vulnerabilities are of many kinds. U.S. power projection forces might be subjected to an EMP attack by an enemy calculating—mistakenly—that such an attack would not involve risks of U.S. nuclear retaliation. The homeland might be attacked by terrorists or even state actors with an eye to crippling the U.S. economy and American society. From a technical perspective, it is possible that such attacks could have catastrophic consequences . . .

Prior commissions have investigated U.S. vulnerabilities and found little activity under way to address them. Some limited defensive measures have been ordered by the Department of Defense to give some protection to important operational communications.[30]

THOUGH THE U.S. IS HARDLY HARDENING, OTHER NATIONS ARE

Clearly, the Pentagon has not prioritized the threat of EMP, but other countries, such as China, Iran, North Korea, and Russia, have not been so cavalier. They have built offensive EMP attack weaponry into their military doctrine, according to the Congressional Research Service. As far back as 2008, analysts believe, these nations acquired EMP devices that could be used against American assets at a military turning point. As Dr. Wilson wrote, "Several nations, including reported sponsors of terrorism, may currently have a capability to use EMP as a weapon for cyber warfare or cyber terrorism to disrupt communications and other parts of the U.S. critical infrastructure. Also, some equipment and weapons used by the U.S. military may be vulnerable to effects of EMP."[31]

This assessment is widely shared by defense analysts in the private sector, and within the Pentagon itself. In 2010, the aforementioned Dr. Ashton Carter did a survey of the military's preparedness for nuclear and EMP attacks. He concluded that some progress had been made in recent years, but that major weapons systems designed to protect the nation are still inadequate. Specifically, he reported the following:[32]

- The Army has "an improved process for independent review of survivability" which sounds like so many bureaucrats shuffling deck chairs.

- The Navy has "implemented a requirements review process," which sounds a lot like the Army's response.

- U.S. Strategic Command (USSTRATCOM) has "continued to devote resources and talent to identifying mission critical capabilities and assessing their survivability," which also sounds like the Army, only with rhetorical flourish.

- The Air Force has "committed resources" to test major platforms for high-altitude EMP protection, which sounds slightly more proactive.

Dr. Carter's report was prepared for the Defense Science Board, at the behest of Congress and the DoD, to monitor the military's progress in preparing for EMP. The impetus for the report was the concern in Congress that after the EMP Commission officially disbanded, the threat of EMP would be become a nonpriority.

Which is what happened.

After the EMP Commission was disbanded, several members of Congress sent petitions to the DoD, urging continued vigilance and preparation. The DoD replied that it supported the commission's findings in concept, but felt those findings would be best addressed by folks over at the Department of Homeland Security (DHS). Apparently, bureaucratic turf wars would take precedence over national security. And now, despite Dr. Carter's blunt appraisal of military unpreparedness, little has changed:

- The Army, Navy, and Strategic Command continue to think about thinking about the problem.

- The Air Force has actually taken action, though it has yet to implement EMP protections on the new stealth F-35, tankers, next-generation bombers, *Air Force One*, or other vital aircraft.

- Our military forces still positioned in the Middle East and Central Asia remain defenseless against an EMP attack.[33]

MISSILE DEFENSE AGENCY NOT ONBOARD

More than twenty-five years ago, President Reagan challenged the U.S. scientific community to develop antiballistic missile technologies that would improve our national security and reduce our reliance on nuclear weapons. Today, the Missile Defense Agency (MDA) is doing just that. America's missile defenses are the world's most impressive, and intimidating to any would-be aggressor, keeping us safe from our enemies on every level.

Probably.

The MDA has not been forthcoming in reporting on its own efforts to harden against EMP attack. It claims to be operating under different criteria for hardening, without saying what those criteria are or whether any steps have been taken to implement the hardening. MDA's rather vague and tight-lipped public position may be purposeful—to keep potential aggressors in the dark about our true deterrent capabilities. But this vagueness has been sufficiently worrying to Dr. Carter that he joined other EMP analysts in reporting their concerns: [34]

- There is an "overall fragmentation of efforts—little movement to a national enterprise as recommended by two previous DSB [Defense Science Board] task forces."

- Despite recommendations of two EMP Commissions and independent analysts answering directly to the secretary of defense, EMP is not considered an issue by the DoD, and no concerted effort is being made to address the issue head-on.

- The DoD's Defense Threat Reduction Agency, which was created in 1998 as the "Combat Support Agency for countering weapons of mass destruction," does not see EMP as a priority issue.

This last observation about the Defense Threat Reduction Agency is the oddest. Its focus is solely on WMDs, which it correctly defines as including biological, chemical, and nuclear weapons, as well as delivery systems. And yet this agency has written:

While a blast of radiation might not do anything to properly protected troops, it would "fry" anything electronic: laptops, sensors, our highly computerized planes, even a simple cell phone. A weapon that doesn't kill a single person could still destroy our technology. To counter that threat, DTRA developed Radiation-Hardened technology for the hardening of microelectronic technology against radiation effects. In a harsh radiation environment our warfighters must be able to move, shoot and communicate if they are to defend the United States and our allies and prevent additional use of WMD—DTRA is making sure they can do just that.[35]

From this passage, you may think the agency is fully on board with the EMP Commission. But Dr. Carter's report says the agency has shown "little progress toward a 21st century approach."[36]

A REMARKABLE VULNERABILITY—THE PENTAGON USES THE SAME LINES YOU DO

Both the EMP Commission and the Defense Science Board pointed to one particularly glaring weakness: our military is highly reliant on commercial communication lines. If the military used Verizon and AT&T phone lines only occasionally, this would not be a cause for concern. But according to the independent commission, the Pentagon uses commercial infrastructures, particularly the electrical grid and communications and navigation systems, in up to 99 percent of its operations in the U.S. So if the civilian grid is knocked out of action, the Pentagon will be severely crippled as well.

Many people are surprised to learn just how dependent out military is, and would be in a war on American soil, on civilian resources. In recent interviews with local officials in Rapid City, South Dakota, for example, these sources reinforced the fact that military facilities in the area, especially those with missiles, rely totally on the civilian economy for their electricity and such critical utilities as water.

Prior to his death in September 2000, retired U.S. Army lieutenant general Robert L. Schweitzer pointed out:

Ninety percent of our military communications now passes over civilian networks. If an electromagnetic pulse takes out the telephone systems, we are in deep trouble because our military and non-military nets are virtually inseparable. It is almost equally impossible to distinguish between the U.S. national telecommunication network and the global one . . . It is finally becoming possible to do what Sun Tzu wrote about 2000 years ago: to conquer an enemy without fighting. The paradigm of war may well be changing. If you can take out the civilian economic infrastructure of a nation, then that nation in addition to not being able to function internally cannot deploy its military by air or sea, or supply them with any real effectiveness—if at all.[37]

Even now, that dependency remains unacceptably high.

This "ubiquitous dependence" on commercial-off-the-shelf (COTS) technology in almost all military-supported operations can lead to serious problems, says the Defense Science Board. It "creates a twofold downside when considering nuclear survivability."[38]

By "twofold downside," the board means that (1) the military has no way of knowing the impact a nuclear EMP attack will have on any COTS device in a military situation, and (2) there is no way to test whether commercial technologies can reach back to obtain the materiel that forward-deployed troops may need in an major EMP event. In short, the DoD doesn't truly know the near term or the long-term effects EMP will have on our fighting forces.[39]

Of equal concern to Dr. Carter and the EMP analysts is the DoD's nuclear weapons expertise. That expertise has decayed significantly in recent years. There were at least fifteen specialists at the Pentagon focusing on nuclear weaponry during the Cold War period. Since then, those specialists have retired or relocated in private industry. Dr. Carter did not mince words in his assessment of our nuclear preparedness: "[If] subjected to a nuclear event in the foreseeable future, mission execution would depend upon combinations of luck and ingenuity in workarounds for failed equipment. There would almost certainly be an unnecessarily high human cost." [40]

Further evidence of the Pentagon's lukewarm concern for EMP came from a recent "Nuclear Posture Review." That review did not address nuclear survivability, except indirectly, in terms of strategic deterrence. It did not address the vulnerabilities of our military systems to EMP effects. And the reason for this? Its authors said, "This area is not one on the 'radar screen' for conventional operations."[41]

THIS INFORMATION IS *NOT* CLASSIFIED

When you read how vulnerable we are at every level from the military high command down to the local reservoir, you have to wonder, *What are our leaders thinking?* The Pentagon has to be aware of these vulnerabilities at every level of the organization. There are no high-level classifications on this information, keeping it away from any key decision makers or oversight commissions. That is obvious—we have this intelligence! And we have publicly reported evidence that EMP and RF devices are being used in warfare right now. Our nation is using them. Other nations are too. So are jewelry thieves in Amsterdam. Yet very little is being done at the Pentagon to protect our nation from a wholesale EMP attack.

In the National Defense Authorization Act for Fiscal Year 2011, the House Armed Services Committee made it clear that it is concerned about an attack. Congress directed the Pentagon to contract with outside experts to "conduct an assessment of Department of Defense plans for defending the territory of the United States against the threat of attack by ballistic missiles, including electromagnetic pulse attacks."[42] Once the Defense Department completes this assessment, it is then supposed to be reviewed by the Government Accounting Office (GAO) to make sure that taxpayer money is being well spent.

This very process has occurred multiple times over the last decade. A few concerned members of Congress succeeded in prompting a bare minimum of action from the brass. Military analysts are ordered to study the situation with all the rigor they are capable of. Months later they produce a report recommending that the matter be studied further, or that limited test projects be initiated. They report as much to the GAO, which concurs wholeheartedly. Nothing changes, and the nation remains unprepared for an attack even though retired general Eugene Habiger, former commander in chief of the U.S. Strategic Command, has issued a most dire warning . . .

"It is not a matter of if, it is a matter of when."[43]

8

WE CAN STILL DO THE
RIGHT THING

SIXTY YEARS AFTER the first nuclear detonations warned U.S. leaders that electromagnetic pulses could become the asymmetric game changer in future warfare, the entire issue sits bogged down in bureaucratic indifference and political scapegoating.

When the 2008 EMP Commission assessed the potential for a hostile nation or terrorist group to attack the United States with a high-altitude EMP weapon, the commissioners determined that any number of adversaries possess both the ballistic missiles and nuclear weapons capabilities and could in fact perform a high-altitude EMP attack against the U.S. within the next fifteen years. Furthermore, North Korea, Pakistan, India, China, and Russia are all in the position to launch an EMP attack against the United States now.

To improve U.S. preparedness, the EMP Commission addressed ten critical areas of the national infrastructure and offered seventy-five unclassified recommendations. The commission called upon the Department of Homeland Security to make clear its responsibility for responding to an EMP attack and for developing contingency plans in cooperation with federal, state, and local agencies, as well as industry.

Yet little action has been taken.

EMBARRASSINGLY LITTLE HAS BEEN DONE

According to subject matter experts Carafano, Spring, and Weitz, who have been referenced often, Homeland Security in its larger capacity has identified fifteen threat scenarios to prepare for on a national scale. These are what they call "high-consequence threat scenarios, such as terrorist attacks or natural disasters."[1] But an EMP attack is not included among them.

What's more, the 2008 National Defense Authorization Act specifically directed Homeland Security to coordinate with the EMP Commission to protect against just such an attack. Forced by the congressional directive to do *something*, the bureaucrats at Homeland Security held a number of meetings to decide when to hold future meetings to determine agendas to be acted upon at yet future meetings. And into the bureaucratic miasma the order and intentions sank. And so the experts concluded: "Despite the grave dangers posed by an EMP attack, an EMP threat scenario has yet to be incorporated into [Homeland Security's] National Planning Scenarios."[2]

Since then, Homeland Security has addressed EMP only once, at a June 2010 Critical Infrastructure Partnership Advisory Council Joint Sector meeting (that title chosen after many meetings, no doubt). At the meeting, there was a ten-minute "Electromagnetic Pulse Update" from 2:40 to 2:50 p.m.[3]

In the next meeting of the council, held a month later, there was no mention at all.

This indifference is not limited to Homeland Security and the Pentagon. Experts Carafano, Spring, and Weitz say there's "no energy at the Department of Energy" either for dealing with EMP. Though the Energy Department received $11 billion in funding to modernize the grid, little has been done to protect against an EMP attack or natural occurrence. They have "tentatively begun to identify and take appropriate corrective action" to protect the U.S. power system. But like Homeland Security and the Pentagon, they have barely moved past the theorizing and the meeting taking to protect the electric grid on which we rely.[4]

We gained a better insight into the reason for the foot-dragging from a July 2009 workshop cohosted by the Department of Energy and the North American Electric Reliability Corporation (NERC) in which the agenda was "high-impact, low-frequency" events, such as EMP.[5]

NERC is an organization of electrical grid operators formed to "help maintain and improve the reliability of North America's bulk power system" and "attain operational excellence" by "identifying issues before they have a chance to become critical."[6] NERC's business is entirely dependent on a fully functioning electrical grid, so they are keen to work with the Department of Energy and others at the workshop. There were, in fact, 110 others in attendance, including representatives from the Departments of Energy, Homeland Security, Defense, and Health and Human Services; the EMP Commission; and private utilities.

Among the threats examined were natural geomagnetic disturbances and electromagnetic pulses, in addition to dangers from a cyber attack on the U.S. energy infrastructure, and any ensuing pandemic affecting millions of Americans in a sudden crisis situation. At the conclusion of this workshop of government and industry leaders, a report was issued. These excerpts relate to the EMP threat:

> Collective action is needed to reconcile real and valid concerns about cost, labor and the sector's shrinking workforce with the legitimate questions of national security posed by coordinated physical and cyber-attacks and High-Altitude Electromagnetic Pulse, or HEMP, weapons . . .
>
> As the military landscape has changed from nation-state threats to a greater concern over terrorist and rogue-nation threats, the risk for the use of weapons of mass destruction (such as EMP weapons) has increased . . . Adversaries are more likely to resort to asymmetric means, using unconventional approaches that avoid or undermine North America's strengths and instead exploit vulnerabilities. This means that the future target of HEMP may well be the civil infrastructure of the United States as opposed to military systems, which have considered the HEMP threat for many years.[7]

In reading these excerpts and other reports offered by the workshop participants, you could conclude that the participants see value in implementing the recommendations of the EMP Commission. They believe government and industry should be finalizing plans to reduce the time needed to restore power following an attack, whether from deliberate or natural means. And a "defense plan" should be drawn up that details system design specifications, hardening strategies, and power restoration methods.

Two years later, the workshop participants have taken only baby steps.

The industry classifies the EMP risk as "low-probability, high consequence," which means they know things will be horrible if the bomb goes off, but they're betting—with your community, home, and family at stake—that nothing is going to happen. They do not even invest in surplus inventory of large transformers or support the manufacturing of the transformers in the United States to ensure availability. The R&D (Research & Development) in the energy industry in the U.S. is reported to be 0.3 percent. Among all industries, R&D runs at 3 percent, ten times higher.[8] The interest in dealing with EMP couldn't be lower.

WHAT IS BEING DONE TO PROTECT THE GRID?

As uninterested as the energy industry is in confronting this problem, there has been a small bit of movement.

WAUKESHA ELECTRIC

In 2010, at least one major company that makes small- and medium-sized power transformers, SPX Corp, joined with the Waukesha Electric power company to build a new manufacturing plant dedicated to outputting large transformers. SPX chief executive Christopher J. Kearney said, "The expansion of our SPX Transformer Solutions facility in Waukesha is an important milestone and underscores our long-term commitment to manufacturing medium and large power transformers for our valued power transmission and distribution customers throughout North America. By expanding our facility, we have significantly increased our capacity to produce the transformers that we believe will be an essential component of addressing America's rapidly aging power grid infrastructure."[9]

MITSUBISHI ELECTRIC

In February 2011, Mitsubishi Electric announced plans to begin building large power transformers by early 2013 in a new, $200 million, 350,000-square-foot facility occupying almost one hundred acres in the Rivergate Industrial Park in Memphis, Tennessee.[10]

EDISON ELECTRIC INSTITUTE

Ed Legge of the Edison Electric Institute in Washington says, "We're taking this seriously" because the studies done together with NERC and the Energy Department suggest that threats to the grid "may be much greater than anticipated."[11]

But beyond these examples of corporate responsibility and commercial risk-taking, there is little interest in the energy industry of making any changes, particularly ones that could be expensive. In fact, the little interest that had existed, that had driven policy makers and private industry, has subsequently been squashed . . .

AT THE CRUX OF THE PROBLEM

In 2009, NERC went under new management. The new chief executive officer, Gerry Cauley, recently testified to the Senate Energy and Natural Resources Committee, and according to reporting from Dr. Pry, insisted that "geomagnetic storms cannot threaten transformers, the operation of the electric grid or the survival of the American people."[12]

NERC's official position has clearly evolved. Now they are maintaining that a geomagnetic super storm like the 1859 event would *not* cause widespread damage to big transformers or a catastrophic collapse of the national electric grid for a protracted period. So the NERC is now at odds with reports by the 2008 EMP Commission, the Department of Energy, the Federal Energy Regulatory Commission, and national laboratories at Los Alamos, Lawrence Livermore, Sandia, and Oak Ridge. Reports from all these government groups concluded that a great geomagnetic storm or EMP event could damage and destroy hundreds of electrical transformers and take down the national electric grid for three to ten years. That, in turn, as Dr. Pry has told us, could cause "millions of Americans to perish from starvation and societal collapse."[13]

Similar concerns have been raised across the pond. In February 2012, the British Parliament's Defense Committee did its own independent investigation of EMP and geomagnetic storms. The committee, which is the equivalent of our Senate Armed Services Committee, called for greater protection of the UK grid from geomagnetic storms and other electromagnetic pulse threats. So Britain's political leadership and its scientific and technical community agree with the U.S. studies warning about the catastrophic threat and are urging the hardening of the electric grid.

NERC stands alone against this international consensus.

THE EXPERTS WERE SHUT UP

Even a number of the participants of NERC's own Geomagnetic Disturbance Task Force, according to Dr. Pry, said the report they wrote was designed to misinform policymakers so that the electric power industry would not be forced to invest in hardening the national electric grid. Task force participants identified a number of omissions and misrepresentations in their own report:

- NERC's final report focused only on geomagnetic storms and excluded data on the effects of nuclear and non-nuclear electromagnetic pulse threats—for which there is more experience and empirical evidence.

- This same report also deferred to the opinions of the transformer manufacturers—who would be understandably loath to detail how their products might fail. The more impartial views of EMP experts were not included in the official report.

- The report did not include data on the vulnerability of transformers and actually excluded incident cases where transformers were known to have been damaged by geomagnetic storms. They tried to create the fiction that in documented cases of damaged transformers, the damage resulted from other causes.

- Prior to publishing the final report, NERC did not, as is customary, circulate a draft or investigate any studies that contradicted its own conclusions. The report was instead written in secret and was not made available for critical review. Despite a public protest letter from the NERC's own technical experts, NERC went ahead with its publication.[14]

In the ten-page detailed paper from the task force that sought to make public these concerns, the experts who actually understood the issue made it clear that business concerns were taking precedence over science. The financial bottom line of U.S. power companies was prevailing over national security. And given that the electrical grid can be made secure at a relatively low cost, and it's not being done, it is no longer such an impressive wonder. It is a clear and present danger on native soil.

The task force reiterated a key point: It isn't just the instantaneous failures that will occur in an EMP attack or geomagnetic storm. It's the fact that downed transformers could take years to replace, and the resultant effects on every local economy would be devastating.[15] We return to the technical study done by the Sage Group on the "grid down" impacts we would face:

> There are likely to be many constraints on the ability of the region's economy to restore itself. For example, between 1,100 and 1,200 medium to large power transformers are sold in the United States each year. The largest transformers in the region take one to two years to build and additional time for deployment. The best financed utilities strive to keep as many as 10 percent of their transformers in stock since their schedules assume as long as five years to put one into use. If an EMP resulted in an instantaneous demand for 10–70 percent of replaced medium to large power transformers, it would require using all of the spares (assuming the spares were not damaged) at a minimum and 600 percent more in the worst case. It is unclear how quickly replacement transformers could be delivered to the region. Once what little stock available is used, customers would have to wait in line for their turn to receive their orders. Additional build-out of factory facilities would likely need to be made to fulfill large back logs of orders. At that point, we would hope that in the scenario presenting the least damage, there would not be a second EMP hit six months later when no spares are left.
>
> Skilled labor would also be an essential part of a restoration effort. The mobilization of utility workers in response to natural disasters is a familiar story. An EMP, however, would likely require similar mobilizations of skilled personnel to restore the electrical and electronic infrastructure. It is unclear whether the industries and businesses required for infrastructure restoration have the organizational experience or capacity to quickly mobilize in response to an EMP disaster or to sustain the effort needed for recovery.
>
> The financial investment necessary to restore economic capacity on a regional basis is substantial. Presumably the insurance industry

would be a critical player in amassing the capital necessary for this investment. The experience of Hurricane Katrina, however, suggests that the insurance industry may at times be a barrier to recovery. ✔ Were an EMP considered an act of war, insurance claims might be legitimately rejected. For a variety of reasons, the lack of capital could substantially delay the restoration of economic capacity and increase the economic costs associated with an EMP.[16]

This is surely an ominous analysis, but the Sage analysts also offered cause for hope.

A WAY TO BEGIN DOING THE RIGHT THING

The Sage report said that if a community could protect the vital 10 percent of its critical infrastructure, such as energy and communication, it could gain much more than a 10 percent reduction of loss. "Because partial shielding will reduce damage to critical infrastructure, it will tend to reduce the time required to recover full economic vitality."[17]

And so, if a metro area retrofitted roughly 10 percent of its strategic infrastructure with EMP-protected micro-grids and related systems and emergency communications through the use of shielded systems and related SCADA devices, it could ensure its water supply, the functioning of its vital emergency communication centers, and its hospitals. In the days following a catastrophic event, this "10 percent edge" just might make the difference.

We can begin with a 10 percent difference. That is the reasonable thing to do, the urgent action we should take, the message we should send to our elected leaders who will listen.

But if they do not . . .

9

ONE SECOND AFTER, LITERALLY

Suppose the old America, so wonderful, the country we so loved, suppose at four fifty P.M. eighteen days ago, it died. It died from complacency, from blindness, from not being willing to face the harsh realities of the world. Died from smug self-centeredness. Suppose America died that day.

—From William Forstchen's *One Second After*[1]

SURVIVING IN FREEDOM—A PERSONAL MANIFESTO

As we've seen, when an EMP attack does come, we won't be able to rely even on that exasperating old line "I'm from the government, and I'm here to help." We cannot expect a thing from Washington. In fact, if the nation's capital is hit, they will be in worse shape than people in the Dakotas, the Carolinas, and even the desert states. As depicted in the chilling cautionary tale by William Forstchen, *One Second After*, each of us in our communities will be on our own. And as a whole, America will be a nation forsaken.

So the first step in preparing for an EMP event is to deal with reality: the federal government will not be there to help you in any meaningful way. You're on your own. Your world will never be the same again. But if you have prepared—if you have taken the common-sense steps we outline here—you and your family and your local community will soldier through the worst of the days ahead.

To begin helping to alleviate this shock, local leaders at the community level could begin now to prepare for an eventual and possibly inevitable EMP event. For example, they can have in place a plan for area residents to report to pre-positioned centers where food, water, generators with fuel for lighting, and basic medical and pharmaceutical supplies would already be stocked to meet emergency needs.

Such centers would be set up at strategic locations to accommodate citizens who may be coming in from small towns or outlying farm areas. Emergency first responders as well as law enforcement would have a procedure established to report directly to these centers in the event of such a crisis. Such centers also would be useful in keeping citizens informed of the latest developments.

Assuming that there would be no communications, local leaders would already have distributed flyers to each household to inform residents where these centers would be located.

Such centers also should be equipped to handle small tent cities that would be set up for those citizens who have to travel long distances to reach these pre-positioned centers.

Such an approach assumes that local leaders would take the initiative for their respective jurisdictions and not wait until it's too late expecting state or federal assistance to arrive in a timely fashion.

Then, there are steps that can be taken for yourself and your family.

Your first step, again, is to assume the cavalry won't come. So your second step is to prepare a "go-bag" so you're always ready to go on a moment's notice. This go-bag should be a large, lightweight, nylon backpack that's well fitting—one for each member of the family, and for the pets, as well. Into this go-bag you should pack the bare essentials needed for two to three days of survival in the outdoors. If the EMP event is confined to a limited surrounding area, you may be able to move to safe ground in that many days.

30 ITEMS TO PACK IN YOUR GO-BAG

Each of these items could become "essential" and has been listed in descending order of importance to survival in an emergency EMP event or geomagnetic storm.

1. Prescription medications, backup quantities

2. Water—enough for 3 days, plus water purification tablets

3. High-protein food—freeze-dried, enough for 6 meals

4. ID card and passport (have photocopies packed in case originals are not at hand)

5. First aid kit—high quality—include Cipro should you consume bad food and water, charcoal in event of poisoning and antihistamine for allergies.

6. Cash—a few hundred dollars at least (banks will be closed, ATMs not functioning)

7. Hiking shoes—premium quality

8. Warm clothing—one complete set

9. Backup pairs of prescription eyeglasses or contact lenses, and sunglasses

10. Battery-operated radio, with extra batteries

11. LED flashlight (requires no bulbs) and extra lithium batteries

12. Multi-tool (recommended: Swiss Tool Spirit Plus) for cutting, opening cans, punching holes, screw driving)

13. Six-inch fixed-blade camping knife—include a spoon and fork or a "spork"

14. Pocket knife, up to 4 inches

15. Machete (recommended: 5/16-inch Cold Steel Gurka Kukri in SK-5 high carbon), very practical for bringing down trees

16. Small handsaw for cutting branches to make a fire

17. Knife sharpener (recommended: a handheld tungsten carbide sharpener from Lansky, for example), to put a fine edge on your tools

18. Fire starter/strike-anywhere matches (recommended: Blast Match by Ultimate Survival—designed to be used with one hand in the event of an injury)

19. Vaseline—excellent as a fire starter and in numerous first-aid situations

20. Compass and a paper map of your region

21. Fishing line—50 feet, 15-pound test—useful for sewing, suturing wounds, laying traps to catch small prey to eat

22. Rope—25 feet, to tie and secure items

23. Guns—one good revolver or semi-automatic with ammo at a minimum; rifle and shotgun useful

24. Surgical mask (to protect against germs) or scarf to cover your face in foul weather

25. Gloves, to protect your hands, guard against cuts (infection is your biggest initial concern), and keep you warm

26. Roll of duct tape (because there just isn't anything it can't do)

27. Small sewing kit for mending clothes or closing up wounds

28. Hammocks, with a cheap tarp that can be strung over the top, or a tent if you prefer—be sure the fabric is high-tensile lightweight nylon

29. Space blanket

30. Gold—a few ounces for insurance or barter in a desperate situation

These are the primary items for your go-bag. Obviously your backpack will be full to overflowing with all of these items. If you can divide up the hard goods among your family's backpacks, you will be set.

PREPARING FOR THE LONG ORDEAL

Your third step is to begin marshaling your resources. There are eight things to begin thinking about, at a manageable pace, in the months ahead. Listed here are questions to prompt your thinking, steps you can take in an orderly fashion, and suggestions for approaching your particular local situation.

1. WATER
This is the most important resource of all. Securing a dependable source of pure water is your top priority.

- Take stock of current water sources in your area, and how likely they are to be impacted:
    1. Natural sources (creeks, rivers, year-round springs?)
    2. Nearby wells (powered by electricity, manual, solar?)
    3. Storage basins (ponds, tanks, water catch systems, barrels)

- How will you capture water on your own property—house downspouts into barrels? basins? a swimming pool? a well?

- Obtain 5-gallon plastic buckets for transporting water (any larger and you risk back strain).

- Obtain a two-wheel garden cart for hauling water (among other things).

- Obtain plain Clorox bleach or iodine to treat water (1/4 teaspoon purifies one gallon).

- Obtain a ceramic water filter

2. FOOD

It's a daunting proposition, but to be fully prepared you will want to stockpile a full year's supply of nonperishable food.

- Begin by buying extra quantities of canned products that you normally use, and place the extras in storage.

- Obtain a week's worth of military-specification Meals, Ready To Eat (MRE)

- Obtain large (5-gallon) food-grade buckets of wheat, rice, beans, sugar, and salt.

- Expect other family members and friends to become long-term houseguests when they learn that you've planned ahead and they did not. And charity does begin at home. So plan to cook for larger numbers, beginning with a restaurant-sized frying pan, stew pot, and kettle—because you will probably be heating water in the fireplace or wood-burning stove for bathing, dish washing, and laundry.

- Learn to build your own solar oven (for baking hearty and delicious bread).

Post-EMP, farmers and gardeners with green thumbs become kings. Here's what you need to do:

- Stock in a full supply of gardening tools, and spares.

- Obtain fencing against deer (tall), rabbits (fine mesh), and gophers (deeply planted).

- Obtain long-term-storage seeds—select nonhybrid heirloom seeds that are hardy and suitable for growing in your climate. Also obtain topsoil amendments and fertilizers.

- Begin growing medicinal herbs (such as aloe vera for burns) that could become lifesavers when medical supplies run low.

If you are not already an outdoorsman, begin taking a hobbyist's interest in foraging—from hunting and fishing to trapping. Then take these steps as well:

- Obtain night-vision goggles with a maintenance kit and battery charger. Animals come out at night, more so when they're being hunted en masse.

- Take a tip from James Wesley Rawles, author of *How to Survive The End of the World as We Know It*, and *don't* go out to hunt. It takes too much effort, now that effort matters. Lure the game to you with salt blocks. Store up a bunch.

- Obtain some small-game snares.

- Fly-fishing is fun, but invest in practical spin-casting gear and tackle—to increase your odds of success—handmade fishing poles with line, hook, and sinker still work.

- Obtain some frog gigs, crawfish traps, or fish traps—whichever of these is more abundant in your area.

- Make an inventory of all your power tools and begin obtaining hand tools that accomplish much the same ends. You should have tools for gardening, auto repair, welding, woodworking, gunsmithing, and grain grinding, at the very least.

3. HEALTH

In a highly prosperous America laden with every comfort, it's easy to become physically weak, soft, and lazy—without consequence. But after an EMP attack, you need to be physically stronger, less dependent on medications, in better overall shape.

- This is a wake-up call to begin reintroducing a fitness regime, working off that gut, getting in shape, strengthening the lower back—for the more demanding physical tasks required for self-sufficiency.

- Also important is keeping your dentistry current and completing any elective surgery you've been contemplating.

- Make a list of items you now rely on to relax you in stressful situations—such as books, games, liquor, chocolate—and store some away. Forget about the CDs and MP3 players—batteries to power these will quickly become a precious commodity. But old-fashioned musical instruments make good sense.

- Plan on digging and rigging for an outhouse, and begin stacking in bags of powdered lime and toilet paper in quantity.

- Obtain bottled lye to make soap (which may become the ultimate luxury).

You could go many months without power, using fire to cook, axes to chop, guns to hunt and guard the homestead, old machinery to harvest the crop—though you're unfamiliar with it. And all without properly functioning hospitals. So the watchword is CAUTION! But also prepare for injuries small and large.

- Obtain first-aid treatments for burns, cuts, wounds, and insect bites.

- Obtain a minor-surgery outfit with stainless steel instruments. You may have to learn how to use them, or put them in the hands of someone with experience.

- Your go-bag should contain all the essential personal items on which your family depends. But also make a separate list of items needed long-term by each family member and pet. Items to obtain include medications, eyeglasses and contacts, sunscreen, birth control, feminine-hygiene products, toothpaste, and floss.

Within a month of an EMP attack, your top concern will shift from having enough food and water to avoiding influenza or other contagious diseases spreading through the area. You need to be able to quarantine the sick in your area, or short of that, seal off your family from infection.

- Obtain rolls of plastic sheeting and duct tape to cover all the doors, windows, and openings in your quarantine area. Also obtain HEPA filters to purify incoming air.

• Look into installing an outdoor shower just outside your safe house.

• For complete protection against EMP and other potential disasters, also stock up on potassium iodate (KI03) tablets, N95 respirator masks, a steam vaporizer, antibiotic medications, and disinfectants.

4. ENERGY

When the lights go out, people go nuts. You don't want to be the only family in the area with lights shining brightly at night.

• Obtain one or two good generators, with plenty of extra fuel and oil.

• Obtain battery-charging devices—there are good solar options and unconventional options such as attaching a stationary bike with a belt to an alternator to a battery to an inverter to an outlet; you get electricity and a workout! Or you can attach it to a grain grinder to turn wheat into flour.

• Obtain materials for making blackout screens for your windows.

• If you are fortunate (post-EMP) to have propane-fueled appliances, obtain a "tri-fuel" generator that handles gasoline, propane, and natural gas.

• Obtain kerosene lamps with plenty of extra wicks, mantels, and chimneys, as well as ample supplies of fuel.

• Lay in several cords of firewood.

• Obtain the largest propane, heating-oil, gas, or diesel tanks that you (1) can afford and (2) are permitted by local ordinances. Keep them as full as possible at all times. If your local water table allows it, bury your tanks or locate them out of sight.

• Obtain a good-quality chain saw, such as Stihl or Husqvarna, along with extra chains, critical spare parts, and two-cycle oil.

- Assume your strength will be tested to the point of bone tired. Operating a chain saw may become more difficult than now. So don't take chances. Obtain Kevlar chain saw chaps (there won't likely be an emergency room to visit). Plan to also wear sturdy leather gloves, goggles, earmuffs, and helmet.

5. SECURITY

Post-EMP, you'll want a security perimeter around your home or shelter. You'll want to know your neighbors well, since there is safety in numbers. However, a smaller group of well-prepared and well-trained allies can be an important security asset.

- Look into installing the strongest locks, alarm systems, fences, and cable barriers you can afford, along with landscaping features and obstacles to discourage potential marauders and convince them to move on to easier targets.

- Make a map of your area, with your property in the center. Plot out and memorize all roads, trails, and waterways that could be used as escape routes for yourself, or attack routes for outsiders. Write in the names of your neighbors, along with phone numbers and e-mail addresses (in case communications are still working). Also write in city and county office contact numbers, which may become useful.

- Obtain pistols, as well as mid-range and long-range firearms. Suppressed-sound smaller-caliber weapons are ideal, since loud bangs are an invitation to others to come snooping. Having similar calibers among your weapons makes ammo go farther and work with more than one weapon. This could become important as time goes by. Also have eye and ear protection, cleaning equipment, carrying cases, a reloader, a top-quality scope, spare parts, and gunsmithing tools.

- A longbow or crossbow can be a powerful addition to your weapons arsenal because it is silent, not attracting unwanted attention.

- When buying new clothes, consider moving toward more natural earth tones and sturdier fabrics. They'll be safer and last longer.

- Obtain extra sets of clothing for all seasons—and include foul-weather gear, chest waders, ponchos, gaiters, and mosquito netting.

- Obtain a box of good fire extinguishers—in the immediate aftermath of an EMP event, there will be lots of fires and little if any water to extinguish them.

- A final note on security, again compliments of J. W. Rawles: save your wine corks—burned cork makes quick and cheap face camouflage.

6. COMMUNICATIONS

The most unsettling aspect of an EMP attack may be the possibility of a communications blackout. With the grid down, the air becomes quieter, but hearts will pound when the most connected generation in history is suddenly untethered.

- Obtain at least two radios that run on 12-volt DC power or rechargeable battery packs.

- Store each radio in its own cardboard box; then put that inside a bigger box and wrap it completely with aluminum foil; this creates a basic Faraday cage and protects from EMP.

- Similarly, hide away a pair of FRS/GMRS 2-way radios to ensure valuable 2-way communications after an EMP attack.

- Faraday cages can be built or purchased inexpensively for smaller items, such as radios. Invest in some 8 x 8-inch and 8 x 16-inch bags to EMP-proof and waterproof your sensitive electronics. There are many good suppliers.

7. TRANSPORTATION

Post-EMP, you will be constantly on your feet and moving, but traveling shorter distances. So, as per earlier recommendations, have a great pair of hiking and walking shoes. And brush up on your auto-repair skills, because they will come in very handy.

- Old, carbureted vehicles may be damaged by EMP but can be fixed. So having an old car on the property is a good idea. And install a hitch for a trailer, which will become useful in hauling things and people.

- Obtain a mountain bike if you don't already have one. It may become the best way of getting about and traveling distances, as well.

- ATVs and mopeds are also valuable and require little fuel to perform.

- An inflatable raft will also be useful for crossing rivers and keeping things dry.

8. HOME

If worst comes to worst, your family's chances of survival will depend most on where you live. Having a retreat shelter is crucial; it is also an expensive proposition. So you should begin thinking about your options and how to go about creating the safest haven possible given your resources.

A retreat is not just a cabin in the mountains. It needs to be defensible, like a modern-day castle protecting you from outside danger. Ideally, your shelter will be situated in an area that affords you:

- substantial distance from major urban centers, prisons, and mental hospitals

- a strong agricultural economy and no restrictions on keeping livestock

- considerable distance away from interstate freeways and other easily traveled thoroughfares

- hands-off government, with favorable zoning and building permits

- abundant nearby sources of fuels—gas, wood, and coal

- a hillside position that is easily defensible

- an underground storm shelter and root cellar

- placement upwind from major nuclear-weapons targets

Obviously this is a tall order to fill if you don't already live in such an area. But keep in mind that in a crisis, it's all about the people around you. Too many people in the area means too many potential problems. Too few people means not enough neighbors to pitch in to help one another survive and together rebuild in freedom.

BEING READY WHEN THE EMP STRIKE COMES

You can't know when an EMP disaster will strike. But you can be ready. It's not necessary to go out and empty your bank account to prepare, or to abandon your current lifestyle. Most of the likely scenarios can be survived if you plan for three months without a grocery store, bank, and major utilities. It will take planning to survive. But think of your planning efforts as a life insurance policy—one that pays off with you living instead of the other way around.

Those who teach survival training and emergency preparedness say that living through an EMP disaster involves 80 percent psychology, and 20 percent organization and gear. So in thinking about EMP, it's best to focus on the elements of mental self-reliance and psychological preparedness.

With the right mindset going forward, you can logically attack the long list of physical tasks and challenges. You can begin by actively increasing your own food production capacity—so that you can enjoy true food security come what may. Plant fruit trees and vegetable gardens on your property. If you rent and don't have a space for a garden, then join a community garden. And be sure to buy as often as possible from local growers at the farmers' markets.

There are lots of things you can begin thinking about and taking up as projects. For example:

- Begin canning foods and working toward a goal of growing enough food on your own for your family's survival.

- Learn to cook without electricity or natural gas—a fun project would be to build a solar oven and begin baking nutritious whole grain muffins in it!

- Install a well on a gravity feed or hand pump.

- Install weather-stripping and insulating any places in your home that let in more winter cold than necessary.

- Begin tinkering with Faraday cages—the basic idea is very simple and can be mastered in an afternoon, but in a real EMP or RF attack, complexities enter the picture, and having a working knowledge of electromagnetic shielding will be a valuable skill set.

- Learn how to make alcohol, which will become a valuable liquid for fire starting, sanitizing, cleaning wounds, sterilizing, and of course, drinking.

- Keep bees for making honey and candles, and for trading.

- Begin tinkering with some protective force multipliers—trip wire alerts, motion sensors and noise makers, all battery operated—for the perimeter of your property (Fishing lines attached to rock-filled cans make noise when tripped.)

- Seek training in firearms, military tactics, survival psychology, farming, and hydroponics, as well as certifications in CPR, EMT and paramedics.

A lot of people will tell you it's simply not possible to prepare for something as potentially devastating as an EMP attack. Usually after saying this, they go back to watching their favorite TV show. The truth is, you won't be able to stop an EMP event, but you can prepare to survive it and rebuild. I hope you do that.

ABOUT THE AUTHOR

F. MICHAEL MALOOF, a former senior security policy analyst in the Office of the Secretary of Defense, has almost thirty years of federal service in the U.S. Defense Department and as a specialized trainer for border guards and Special Forces in select countries of the Caucasus and Central Asia. While with the Department of Defense, Maloof was director of technology security operations as head of a ten-person team involved in halting the diversion of militarily critical technologies to countries of national security and proliferation concern and those involved in sponsoring terrorism. His office was the liaison to the intelligence and enforcement community within the Office of the Secretary of Defense in halting diversions and using cases that developed from them as early warnings to decision makers of potential policy issues.

Due to his specialized background, Maloof was assigned under a congressionally mandated Defense Department nonproliferation program to provide specialized training in combat tracking to border guards in Uzbekistan, Kazakhstan, and Kyrgyzstan to deal with the increasing influx of terrorists into those countries. He also provided advanced firearms training to Spetznaz, or Special Forces, of the Interior Ministry of the Republic of Georgia. Additionally, Maloof has provided combat tracking training to U.S. Marines and has a background in executive protection, intermediate and advanced evasive and defensive driving, basic and advanced route analysis, surveillance detection, and attack recognition, as well as off-road driving.

Following the September 11, 2001, terrorist attack on the United States, Maloof was detailed back to report directly to the undersecretary of defense for policy to prepare an analysis of worldwide terrorist networks and to determine their linkages worldwide and their relationship to state sponsors. He then worked on an action plan that constituted the initial Defense Department response to terrorism worldwide.

Prior to his career at the Defense Department, Maloof was a legislative assistant to various U.S. senators, specializing in national security and international affairs.

In between working at the U.S. Senate and the Defense Department, Maloof was a special Washington correspondent for the *Detroit News*, a reporter for a specialized newsletter at *U.S. News & World Report*, and Washington correspondent for the *Union Leader* in Manchester, New Hampshire.

Maloof makes his home in Reston, Virginia, with his family, including their 135-pound red Doberman.

APPENDIX 1

WEBSITES FOR EMP SURVIVAL SUPPLIES

These suppliers are provided as resources—the author and publisher do not have relationships with these companies or necessarily endorse them. Do your due diligence in choosing the suppliers that are best for you in your area.

COMMUNICATIONS

www.icomamerica.com/en/
www.yaesu.com
www.aesham.com
www.arrl.org
www.hamradio.com
www.radioshack.com
www.universal-radio.com

SURVIVAL

www.survivalblog.com
www.thesurvivalistblog.net
www.wilderness-survival.net
www.securityprousa.com
www.americanpreppersnetwork.com

FOOD AND WATER

    www.beprepared.com

    www.foodinsurance.com

    www.ldscatalog.com

    www.mountainhouse.com

    www.nitro-pak.com

    www.providentliving.com

    www.bigberkeywaterfilters.com

    www.xpackprepared.com

GENERAL SUPPLIES

    www.2012supplies.com/d/

    www.cabelas.com

    www.survival-goods.com

    www.uscavalry.com

    www.survivalsolutions.com

    www.survivalunlimited.com

ENERGY

    www.windstreampower.com

    www.harborfreight.com

    www.solarcooker-at-cantinawest.com/

    www.sunforceproducts.com

    www.batterystuff.com

    www.windsun.com

    www.all-battery.com

GOVERNMENT

    www.disasterassistance.gov

    www.fema.gov

    www.ready.gov

    www.whitehouse.gov/issues/homeland-security

APPENDIX 2

RECOMMENDED READING AND SURVIVAL LIBRARY

T HERE ARE A NUMBER OF ADDITIONAL RESOURCES, which outline in greater detail the effects of an electromagnetic pulse event on the Nation's electricity and electronically-based critical infrastructures. What follows are the more important, with references to them for further reading.

REPORT OF THE COMMISSION TO ASSESS THE THREAT TO THE UNITED STATES FROM ELECTROMAGNETIC PULSE (EMP) ATTACK

VOLUME 1: EXECUTIVE REPORT 2004

Dr. John S. Foster, Jr.
Mr. Earl Gjelde
Dr. William R. Graham (Chairman)
Dr. Robert J. Hermann
Mr. Henry (Hank) M. Kluepfel
GEN Richard L. Lawson, USAF (Ret.)
Dr. Gordon K. Soper
Dr. Lowell L. Wood, Jr.
Dr. Joan B. Woodard

OVERVIEW:
EMP IS CAPABLE OF CAUSING CATASTROPHE FOR THE NATION

The high-altitude nuclear weapon-generated electromagnetic pulse (EMP) is one of a small number of threats that has the potential to hold our society seriously at risk and might result in defeat of our military forces.

Briefly, a single nuclear weapon exploded at high altitude above the United States will interact with the Earth's atmosphere, ionosphere, and magnetic field to produce an electromagnetic pulse (EMP) radiating down to the Earth and additionally create electrical currents in the Earth. EMP effects are both direct and indirect. The former are due to electromagnetic "shocking" of electronics and stressing of electrical systems, and the latter arise from the damage that "shocked"—upset, damaged, and destroyed—electronics controls then inflict on the systems in which they are embedded. The indirect effects can be even more severe than the direct effects.

The electromagnetic fields produced by weapons designed and deployed with the intent to produce EMP have a high likelihood of damaging electrical power systems, electronics, and information systems upon which American society depends. Their effects on dependent systems and infrastructures could be sufficient to qualify as catastrophic to the Nation.

Depending on the specific characteristics of the attacks, unprecedented cascading failures of our major infrastructures could result. In that event, a regional or national recovery would be long and difficult and would seriously degrade the safety and overall viability of our Nation. The primary avenues for catastrophic damage to the Nation are through our electric power infrastructure and thence into our telecommunications, energy, and other infrastructures. These, in turn, can seriously impact other important aspects of our Nation's life, including the financial system; means of getting food, water, and medical care to the citizenry; trade; and production of goods and services. The recovery of any one of the key national infrastructures is dependent on the recovery of others. The longer the outage, the more problematic and uncertain the recovery will be. It is possible for the functional outages to become mutually reinforcing until at some point the degradation of infrastructure could have irreversible effects on the country's ability to support its population.

EMP effects from nuclear bursts are not new threats to our nation. The Soviet Union in the past and Russia and other nations today are potentially capable of creating these effects. Historically, this application of nuclear weaponry was mixed with a much larger population of nuclear devices that were the primary source of destruction, and thus EMP as a weapons effect was not the primary focus. Throughout the Cold War, the United States did not try to protect its civilian infrastructure against either the physical or EMP impact of nuclear weapons, and instead depended on deterrence for its safety.

What is different now is that some potential sources of EMP threats are difficult to deter—they can be terrorist groups that have no state identity, have only one or a few weapons, and are motivated to attack the US without regard for their own safety. Rogue states, such as North Korea and Iran, may also be developing the capability to pose an EMP threat to the United States, and may also be unpredictable and difficult to deter.

Certain types of relatively low-yield nuclear weapons can be employed to generate potentially catastrophic EMP effects over wide geographic areas, and designs for variants of such weapons may have been illicitly trafficked for a quarter-century.

China and Russia have considered limited nuclear attack options that, unlike their Cold War plans, employ EMP as the primary or sole means of attack. Indeed, as recently as May 1999, during the NATO bombing of the former Yugoslavia, high-ranking members of the Russian Duma, meeting with a US congressional delegation to discuss the Balkans conflict, raised the specter of a Russian EMP attack that would paralyze the United States.

Another key difference from the past is that the US has developed more than most other nations as a modern society heavily dependent on electronics, telecommunications, energy, information networks, and a rich set of financial and transportation systems that leverage modern technology. This asymmetry is a source of substantial economic, industrial, and societal advantages, but it creates vulnerabilities and critical interdependencies that are potentially disastrous to the United States. Therefore, terrorists or state actors that possess relatively unsophisticated missiles armed with nuclear weapons may well calculate that, instead of destroying a city or military base, they may obtain the greatest political-military utility from one or a few such weapons by using them—or threatening their use—in an EMP attack. The current vulnerability of US critical infrastructures can both invite and reward attack if not corrected; however, correction is feasible and well within the Nation's means and resources to accomplish.

Complete report at: www.empcommission.org/docs/empc_exec_rpt.pdf

REPORT OF THE COMMISSION TO ASSESS THE THREAT TO THE UNITED STATES FROM ELECTROMAGNETIC PULSE (EMP) ATTACK

CRITICAL NATIONAL INFRASTRUCTURES

Commission Members

Dr. John S. Foster, Jr.
Mr. Earl Gjelde
Dr. William R. Graham (Chairman)
Dr. Robert J. Hermann
Mr. Henry (Hank) M. Kluepfel
Gen Richard L. Lawson, USAF (Ret.)
Dr. Gordon K. Soper
Dr. Lowell L. Wood, Jr.
Dr. Joan B. Woodard

April 2008

PREFACE

The physical and social fabric of the United States is sustained by a system of systems; a complex and dynamic network of interlocking and interdependent infrastructures ("critical national infrastructures") whose harmonious functioning enables the myriad actions, transactions, and information flow that undergird the orderly conduct of civil society in this country. The vulnerability of these infrastructures to threats — deliberate, accidental, and acts of nature — is the focus of greatly heightened concern in the current era, a process accelerated by the events of 9/11 and recent hurricanes, including Katrina and Rita.

This report presents the results of the Commission's assessment of the effects of a high altitude electromagnetic pulse (EMP) attack on our critical national infrastructures and provides recommendations for their mitigation. The assessment is informed by analytic and test activities executed under Commission sponsorship, which are discussed in this volume. An earlier executive report, Report of the Commission to Assess the Threat to the United States from Electromagnetic Pulse (EMP) — Volume 1: Executive Report (2004), provided an overview of the subject.

The electromagnetic pulse generated by a high altitude nuclear explosion is one of a small number of threats that can hold our society at risk of catastrophic consequences.

The increasingly pervasive use of electronics of all forms represents the greatest source of vulnerability to attack by EMP. Electronics are used to control, communicate, compute, store, manage, and implement nearly every aspect of United States (U.S.) civilian systems. When a nuclear explosion occurs at high altitude, the EMP signal it produces will cover the wide geographic region within the line of sight of the detonation.1 This broad band, high amplitude EMP, when coupled into sensitive electronics, has the capability to produce widespread and long lasting disruption and damage to the critical infrastructures that underpin the fabric of U.S. society.

Because of the ubiquitous dependence of U.S. society on the electrical power system, its vulnerability to an EMP attack, coupled with the EMP's particular damage mechanisms, creates the possibility of long-term, catastrophic consequences. The implicit invitation to take advantage of this vulnerability, when coupled with increasing proliferation of nuclear weapons and their delivery systems, is a serious concern. A single EMP attack may seriously degrade or shut down a large part of the electric power grid in the geographic area of EMP exposure effectively instantaneously. There is also a possibility of functional collapse of grids beyond the exposed area, as electrical effects propagate from one region to another.

The time required for full recovery of service would depend on both the disruption and damage to the electrical power infrastructure and to other national infrastructures. Larger affected areas and stronger EMP field strengths will prolong the time to recover. Some critical electrical power infrastructure components are no longer manufactured in the

United States, and their acquisition ordinarily requires up to a year of lead time in routine circumstances. Damage to or loss of these components could leave significant parts of the electrical infrastructure out of service for periods measured in months to a year or more.

There is a point in time at which the shortage or exhaustion of sustaining backup systems, including emergency power supplies, batteries, standby fuel supplies, communications, and manpower resources that can be mobilized, coordinated, and dispatched, together lead to a continuing degradation of critical infrastructures for a prolonged period of time.

Electrical power is necessary to support other critical infrastructures, including supply and distribution of water, food, fuel, communications, transport, financial transactions, emergency services, government services, and all other infrastructures supporting the national economy and welfare. Should significant parts of the electrical power infrastructure be lost for any substantial period of time, the Commission believes that the consequences are likely to be catastrophic, and many people may ultimately die for lack of the basic elements necessary to sustain life in dense urban and suburban communities. In fact, the Commission is deeply concerned that such impacts are likely in the event of an EMP attack unless practical steps are taken to provide protection for critical elements of the electric system and for rapid restoration of electric power, particularly to essential services. The recovery plans for the individual infrastructures currently in place essentially assume, at worst, limited upsets to the other infrastructures that are important to their operation. Such plans may be of little or no value in the wake of an EMP attack because of its long-duration effects on all infrastructures that rely on electricity or electronics.

The ability to recover from this situation is an area of great concern. The use of automated control systems has allowed many companies and agencies to operate effectively with small work forces. Thus, while manual control of some systems may be possible, the number of people knowledgeable enough to support manual operations is limited. Repair of physical damage is also constrained by a small work force. Many maintenance crews are sized to perform routine and preventive maintenance of high-reliability equipment. When repair or replacement is required that exceeds routine levels, arrangements are typically in place to augment crews from outside the affected area. However, due to the simultaneous, far-reaching effects from EMP, the anticipated augmenters likely will be occupied in their own areas. Thus, repairs normally requiring weeks of effort may require a much longer time than planned.

The consequences of an EMP event should be prepared for and protected against to the extent it is reasonably possible. Cold War-style deterrence through mutual assured destruction is not likely to be an effective threat against potential protagonists that are either failing states or trans-national groups. Therefore, making preparations to manage the effects of an EMP attack, including understanding what has happened, maintaining situational awareness, having plans in place to recover, challenging and exercising those plans, and reducing vulnerabilities, is critical to reducing the consequences, and thus probability, of attack. The appropriate national-level approach should balance prevention, protection, and recovery.

The Commission requested and received information from a number of Federal agencies and National Laboratories. We received information from the North American Electric Reliability Corporation, the President's National Security Telecommunications Advisory Committee, the National Communications System (since absorbed by the Department of Homeland Security), the Federal Reserve Board, and the Department of Homeland Security. Early in this review it became apparent that only limited EMP vulnerability testing had been accomplished for modern electronic systems and components. To partially remedy this deficit, the Commission sponsored illustrative testing of current systems and infrastructure components. The Commission's view is that the Federal Government does not today have sufficiently robust capabilities for reliably assessing and managing EMP threats.

The United States faces a long-term challenge to maintain technical competence for understanding and managing the effects of nuclear weapons, including EMP. The

Department of Energy and the National Nuclear Security Administration have developed and implemented an extensive Nuclear Weapons Stockpile Stewardship Program over the last decade. However, no comparable effort was initiated to understand the effects that nuclear weapons produce on modern systems. The Commission reviewed current national capabilities to understand and to manage the effects of EMP and concluded that the Country is rapidly losing the technical competence in this area that it needs in the Government, National Laboratories, and Industrial Community.

An EMP attack on the national civilian infrastructures is a serious problem, but one that can be managed by coordinated and focused efforts between industry and government. It is the view of the Commission that managing the adverse impacts of EMP is feasible in terms of time and resources. A serious national commitment to address the threat of an EMP attack can develop a national posture that would significantly reduce the payoff for such an attack and allow the United States to recover in a timely manner if such an attack were to occur.

The complete 208-page report at: http://www.empcommission.org/docs/ A2473-EMP_Commission-7MB.pdf

RECOMMENDED BOOKS

Auerbach, Paul S. Medicine for the Outdoors. Guilford, CT: The Lyons Press, 2003.

Bartmann, Dan, and Dan Fink. Homebrew Wind Power. Masonville, CO: Buckville Publications, 2009.

Black, David S. Living off the Grid: A Simple Guide to Creating and Maintaining a Self-Reliant Supply of Energy, Water, Shelter, and More. New York: Skyhorse Publications, 2008.

Foster, John S., et al. "Critical National Infrastructure: Report of the Commission to Assess the Threat to the United States from Electromagnetic Pulse (EMP) Attack," July 10, 2008, www.hsdl.org.

Miller, Colin R. "Electromagnetic Pulse Threats in 2010." Maxwell AFB, Al: Air War College, Center for Strategy and Technology, November 2005.

Poole, Larry, and Cheryl Poole. EMP Survival: How to Prepare Now and Survive When an Electromagnetic Pulse Destroys Our Power Grid, CreateSpace, 2011.

Rawles, James Wesley. How to Survive the End of the World as We Know It. New York: Penguin Group, 2009.

Skousen, Joel M., and Andrew Skousen. Strategic Relocation—North American Guide to Safe Places. Joel Skousen Designs, 2011.

Stevens, James Talmage. Making the Best of Basics: Family Preparedness Handbook. Gold Leaf Press, WA, 1997.

Thayer, Samuel. The Forager's Harvest: A Guide to Identifying, Harvesting, and Preparing Edible Wild Plants. Ogema, WI: Forager's Harvest, 2006.

HIGHLY RECOMMENDED: Wiseman, John "Lofty". The SAS Survival Handbook, HarperCollins Publishers, 1986.

RECOMMENDED ARTICLES

Statement of Dr. William R. Graham, chairman of the EMP Commission, before the House Armed Services Committee, July 10, 2008 http://www.empcommission.org/docs/GRAHAMtestimony10JULY2008.pdf EMPact America: Prepare, Protect, Recover: Online Resources, Featured Reports http://www.empactamerica.org/featuredreports.php

Before the Lights Go Out: A Survey of EMP Preparedness Reveals Significant Shortfalls, by James Jay Carafano, Ph.D., Baker Spring and Richard Weitz, Ph.D. http://www.heritage.org/research/reports/2011/08/before-the-lights-go-out-a-survey-of-emp-preparedness-reveals-significant-shortfalls

National Academy of Sciences, "Severe Space Weather Events—Understanding Societal and Economic Impacts: A Workshop Report," 2008, National Academies Press, at http://books.nap.edu/openbook.php?booksearch=1&term=emp&record_id=12507&Search+This+Book.x=29&Search+This+Book.y=12

Congressional Commission on the Strategic Posture of the United States, "America's Strategic Posture," United States Institute of Peace Press, 2009, at http://media.usip.org/reports/strat_posture_report.pdf Federal Energy Regulatory Commission, "Electromagnetic Pulse: Effects on the U.S. Power Grid," 2010, at http://www.ferc.gov/industries/electric/indus-act/reliability/cybersecurity/ferc_executive_summary.pdf

Peter Vincent Pry, "What America Needs to Know About EMPs," Foreign Policy, March 17, 2010, at http://www.foreignpolicy.com/articles/2010/03/17/the_truth_about_emps

Kevin Cogan, "In the Dark: Military Planning for a Catastrophic Critical Infrastructure Event," U.S. Army War College, May 2011,at http://www.csl.army.mil/usacsl/publications/InTheDark.pdf

North American Electric Reliability Corporation and the U.S. Department of Energy, "High-Impact, Low-Frequency Event Risk to the North American Bulk Power System," 2009, at http://www.nerc.com/files/HILF.pdf

Applied Physical Electronics, "EMP Suitcase: Compact 2100 Series," 2009, at http://www.apelc.com/applications.html

Michael J. Frankel, testimony before the Judiciary Committee, U.S. Senate, "Government Preparedness and Response to a Terrorist Attack Using Weapons of Mass Destruction," August 4, 2010, at http://kyl. senate.gov/legis_center/subdocs/080410_Frankel.pdf

High-Altitude Electromagnetic Pulse (HEMP): A Threat to Our Way of Life http://www.todaysengineer.org/2007/Sep/HEMP.asp

Directed Energy Weapons and Electromagnetic Bombs, Carlo Kopp http://www.ausairpower.net/dew-ebomb.html

An Introduction to the Technical and Operational Aspects of the Electromagnetic Bomb, Carlo Kopp   http:// www.ausairpower.net/ PDF-A/wp50-draft.pdf

ENDNOTES

CHAPTER 1: A DIRECT ATTACK ON OUR NATION'S CAPITAL

1. D. J. Serafin and D. Dupouy, "Potential IEMI Threats against Civilian Air Traffic," http://www. ursi.org/Proceedings/ProcGA05/pdf/E03.3(0322).pdf, 2–3.
2. Samuel Glasstone, ed., *The Effects of Nuclear Weapons* (United States Atomic Energy Commission, 1962); U.S. House of Representatives; Federation of American Scientists, Intelligence Resource Program, Weapons of Mass Destruction, Intelligence Threat Assessments; Department of the Army, U.S. Army Corps of Engineers.
3. Reggie Beehner, "Simple RF Weapon Can Fry PC Circuits," *PC World*, n.d., http://pcworld. about.net/news/May022001id49048.htm.
4. 1998 Congressional Hearings: Intelligence and Security, Statement of Mr. David Schriner before the Joint Economic Committee, United States Congress, Wednesday, February 25, 1998, "The Design and Fabrication of a Damage Inflicting RF Weapon by 'Back Yard' Methods," http://www.globalsecurity.org/intell/library/congress/1998_hr/s980225ds.htm.
5. "Opening Statement of Lieutenant General Robert L. Schweitzer, United States Army (Ret) to the Joint Economic Committee of the One Hundred Fifth Congress: From 'Economic Espionage, Technology Transfers and National Security' (June 17, 1997, ISBN 0-16-055880-8)," http://blockyourid.com/~gbpprorg/mil/herf/schweit.htm.
6. Information Unlimited, EMP/HERF/Shock Pulse Generators, http://www.amazing1.com/emp.htm.
7. Ibid. http://www.amazing1.com/cgi-bin/mof.cgi?order=1----EMPBG20----Electromagnetic+Pulse+Gun----32000.00.
8. Ibid., http://www.amazing1.com.
9. Ibid., http://www.amazing1.com/download/HAZARDALLFORMS.pdf.

CHAPTER 2: THE EASTERN UNITED STATES—SOMEDAY SOON

1. Testimony of Dr. Michael J. Frankel, Senate Judiciary Committee, Subcommittee on Terrorism and Homeland Security Hearing, "Government Preparedness and Response to a Terrorist Attack Using Weapons of Mass Destruction," August 4, 2010, http://kyl.senate. gov/legis_center/subdocs/080410_Frankel.pdf, 2.

CHAPTER 3: OUR NATION'S CORE VULNERABILITY—
THE POWER GRID

1. James Roberts, "Severe Danger to America of an EMP Attack," http://www.secretsofsurvival. com/survival/emp_attack.html.
2. Steven Aftergood, "Grid Protection and Cybersecurity," *Secrecy News*, June 10, 2010, http:// www.fas.org/blog/secrecy/2010/06/grid_protection.html.
3. Dan Vergano, "One EMP Burst and the World Goes Dark, *USA Today*, October 26, 2010, http:// www.usatoday.com/tech/science/2010-10-26-emp_N.htm.
4. John S. Foster et al., "Report of the Commission to Assess the Threat to the United States from Electromagnetic Pulse (EMT) Attack: Critical National Infrastructures," April 8, 2008, http://empcommission.org/docs/A2473-EMP_Commission-7MB.pdf, 31–32. (Note: Page numbers listed from this report refer to page numbers in the online pdf taskbar rather than those shown on the printed document.)
5. Ibid., 15.
6. Ibid., 21.
7. Charles Perrow, *Normal Accidents: Living with High Risk Technologies*, rev. (Princeton, NJ: Princeton University Press, 1999).
8. The Sage Policy Group, "Initial Economic Assessment of Electromagnetic Pulse (EMP) Impact upon the Baltimore-Washington-Richmond Region," September 10, 2007, http://www. empactamerica.org/10%20EMP%20Econ%20Study.pdf, 6.
9. Ibid., 10.
10. Ibid., 10–11.
11. Ibid., 1, 7.
12. Vergano, "One EMP Burst and the World Goes Dark."
13. Foster et al., "Report of the Commission to Assess the Threat to the United States from Electromagnetic Pulse (EMT) Attack," 22.
14. Ibid., 16.
15. Ibid., 20.
16. William J. Broad, "Nuclear Pulse (III): Playing a Wild Card," *Science*, June 12, 1981, 1248–51.
17. Warren Kozak, "Pearl Harbor, Iran and North Korea," *Wall Street Journal*, December 7, 2011.
18. "Statement, Dr. Peter Vincent Pry, EMP Commission Staff, Before the United States Senate Subcommittee on Terrorism, Technology and Homeland Security, March 8, 2005," http:// kyl.senate.gov/legis_center/subdocs/030805_pry.pdf, 1.
19. Ibid., 3.
20. Clay Wilson, "High Altitude Electromagnetic Pulse (HEMP) and High Power Microwave (HPM) Devices: Threat Assessments" (CRS Report for Congress), August 20, 2004, http:// www.fas.org/man/crs/RL32544.pdf, 11–12.
21. Jerry Emanuelson, "Nuclear Electromagnetic Pulse," *Futurescience*, http://www.futurescience. com/emp.html.
22. "Statement, Dr. Peter Vincent Pry."
23. National Ground Intelligence Center, "(U) China: Medical Research on Bio-Effects of Electro-Magnetic Pulse and High-Power Microwave Radiation," August 17, 2005, available for view and download at http://www.scribd.com/doc/61466831/China-Medical-Research-on-Bio-Effects-of-Electromagnetic-Pulse-and-High-Pwer-Microwave-Radiation.
24. Bill Gertz, "Report: China Building Electromagnetic Pulse Weapons For Use Against U.S. Carriers," *Washington Times*, July 21, 2011, http://www.washingtontimes.com/news/2011/ jul/21/beijing-develops-radiation-weapons/?page=1#.
25. National Ground Intelligence Center, "(U) China."
26. *Wikipedia*, s.v., "Agni-V," http://en.wikipedia.org/wiki/Agni-V#MIRVs.

CHAPTER 4: A DIRECT HIT—FROM THE SUN

1. *Wikipedia*, s.v., "Solar Storm of 1859," http://en.wikipedia.org/wiki/Solar_storm_of_1859.
2. Sten Odenwald, "Chapter 1: A Conflagration of Storms," Space Weather, May 1, 2005, http://www.solar storms.org/SWChapter1.html.
3. Ibid.
4. http://www.nap.edu/openbook.php?record_id=12643&page=5.
5. "New Solar Storm Detector Sending Real-Time Images Used to Warn of Sun's Damaging Storms," *NOAA Magazine*, January 30, 2003, http://www.noaanews.noaa.gov/stories/s1087.htm.
6. Richard A. Lovett, "What If the Biggest Solar Storm on Record Happened Today?" *National Geographic Daily News*, March 2, 2011, http://news.nationalgeographic.com/news/2011/03/110302-solar-flares-sun-storms-earth-danger-carrington-event-science.
7. Tony Philips, "Deep Solar Minimum," NASA Science *Science News*, http://science1.nasa.gov/science-news/science-at-nasa/2009/01apr_deepsolarminimum/.
8. "Plasma Blob From Sun Splashes Earth Today," *Denver Post*, January 22, 2012.
9. Victoria Jaggard, "Magnetic-Shield Cracks Found; Big Solar Storms Expected," *National Geographic News*, December 17, 2008, http://news.nationalgeographic.com/news/2008/12/081217-solar-breaches.html.
10. Associated Press, "Upcoming Sunspot Cycle Likely to Be Strong," *Fox News*, March 7, 2006, http://www.foxnews.com/story/0,2933,187080,00.html. See also the NCAR press release "Scientists Issue Unprecedented Forecast of Next Sunspot Cycle," March 6, 2006, http://www.ucar.edu/news/releases/2006/sunspot.shtml.
11. "HAO 2010 Profiles in Science, Mausumi Dikpati," High Altitude Observatory website, http://www.hao.ucar.edu/Profiles%20In%20Science/Dikpati1.php.
12. Alfred Lambremont Webre, "2012 May Bring the 'Perfect Storm,'" *Seattle Exopolitics Examiner*, April 1, 2009, http://www.bibliotecapleyades.net/esp_2012_53.htm.

CHAPTER 5: THE ROUTE BACK TO THE NINETEENTH CENTURY

1. See Foster et al., "Report of the Commission to Assess the Threat to the United States from Electromagnetic Pulse (EMT) Attack," 131, 201. (See chap. 3, n. 4.)
2. Tim Stevens, "82% of Americans Own Cell Phones," *HuffPost Tech* (*Switched*), November 14, 2007, http://www.switched.com/2007/11/14/82-of-americans-own-cell-phones/.
3. Carlo Kopp, "The Electromagnetic Bomb: A Weapon of Electronic Mass Destruction," http://www.globalsecurity.org/military/library/report/1996/apjemp.htm.
4. Foster et al., "Report of the Commission to Assess the Threat to the United States from Electromagnetic Pulse (EMT) Attack," 100, 103.
5. Ibid., 104–5.
6. Ibid., 100.
7. Ibid., 163.
8. Ibid., 155.
9. Ibid., 155, 158, 159.
10. Ibid., 150.

CHAPTER 6: WHY HAVEN'T WE BEEN TOLD?

1. Statement by Lieutenant General Robert L. Schweitzer, U.S. Army (Retired) before the Joint Economic Committee, United States Congress, June 17, 1997, http://www.fas.org/irp/congress/1997_hr/j970617s.htm.
2. Ibid.
3. kgaines, "American Is Vulnerable to an Electromagnetic Pulse Attack," *Human Events*, June 29, 2009, http://www.humanevents.com/article.php?id=32480.

4. William J. Broad, "Among Gingrich's Passions, a Doomsday Vision," *New York Times*, December 11, 2011, http://www.nytimes.com/2011/12/12/us/politics/gingrichs-electromagnetic-pulse-warning-has-skeptics.html?pagewanted=all

5. See Appendix A in Foster et al., "Report of the Commission to Assess the Threat to the United States from Electromagnetic Pulse (EMT) Attack," 199. (See chap. 3, n. 4.)

6. John S. Foster Jr. et al., "Report of the Commission to Assess the Threat to the United States from Electromagnetic Pulse (EMP) Attack: Volume 1, Executive Report 2004," http://www.militarynewsnetwork.com/reference/empcommission.pdf, 5, 9, 10. (Note: Page numbers listed for this executive report refer to page numbers in the online pdf taskbar rather than the page numbers as shown on the printed document.)

7. Ibid., 11.

8. William R. Graham, Statement before the House Armed Services Committee, http://www.gurevich-publications.com/conspectus/new_congress_warning.pdf, July 10, 2008, 2–3, 5–6.

9. Ibid., 4.

10. Ibid., 2.

11. James Jay Carafano, Baker Spring, and Richard Weitz, "Before the Lights Go Out: A Survey of EMP Preparedness Reveals Significant Shortfalls," *Backgrounder* (published by the Heritage Foundation), no. 2596, August 15, 2011, https://thf_media.s3.amazonaws.com/2011/pdf/bg2596.pdf. This survey is available for download at http://www.heritage.org/research/reports/2011/08/before-the-lights-go-out-a-survey-of-emp-preparedness-reveals-significant-shortfalls.

12. Ibid.

13. Ibid.

14. Ibid.

15. Ibid.

16. http://www.govtrack.us/congress/bills/112/hr668.

17. "EMP Commission," *Right Web*, Institute for Policy Studies, October 19, 2009, http://www.rightweb.irc-online.org/profile/EMP_Commission.

18. Ibid.

19. "Gingrich Warns EMP Greatest Strategic Threat To U.S.," WND Exclusive, November 24, 2011. See http://www.wnd.com/2011/11/370917.

20. Broad, "Among Gingrich's Passions, a Doomsday Vision."

21. http://www.nytimes.com/2011/03/11/opinion/11iht-edholdren11.html.

22. Broad, "Among Gingrich's Passions, a Doomsday Vision."

23. Ibid.

24. Ibid.

25. Ibid.

26. Ibid.

27. Frank J. Gaffney Jr., "The Real EMP Threat: The *New York Times* ignores The Facts to Take Down Newt," *National Review Online*, December 13, 2011, http://www.nationalreview.com/articles/285601/real-emp-threat-frank-j-gaffney-jr.

28. Ibid.

29. "Iran Plans to Send Ships Close to U.S. Waters," September 28, 2011, http://maritimesecurity.asia/free-2/maritime-security-asia/iran-plans-to-send-ships-close-to-us-waters/.

30. "Implementation of the NPT Safeguards Agreement and Relevant Provisions of Security Council Resolutions in the Islamic Republic of Iran: Report by the Director General" (GOV/2011/65), IAEA Board of Governors, November 8, 2011, http://isis-online.org/uploads/isis-reports/documents/IAEA_Iran_8Nov2011.pdf, 8; "UN Nuclear Agency Iaea: Iran 'Studying Nuclear Weapons,'" *BBC News*, November 8, 2011, http://www.bbc.co.uk/news/world-middle-east-15643460.

31. James Reynolds, analysis of "UN Nuclear Agency IAEA" (see previous note); http://www.bbc.co.uk/news/world-middle-east-15643460.

32. "UN Nuclear Agency IAEA."

33. Ibid.

34. Carafano, Spring, and Weitz, "Before the Lights Go Out."

CHAPTER 7: WHAT IS THE CURRENT THREAT ASSESSMENT?

1. Foster et al., "Report of the Commission to Assess the Threat to the United States from Electromagnetic Pulse (EMT) Attack," 22. (See chap. 3, n. 4.)

2. "National Energy Grid Threatened by EMP Attack," TVC Special Report, *Right Side News Daily*, August 18, 2010, http://www.rightsidenews.com/2010081911397/us/homeland-security/national-energy-grid-threatened-by-emp-attack.html.

3. Ibid.

4. Lloyd A. Pritchett, "Outages Stop, but Mystery Remains," *The Sun* (Bremerton, WA), March 27, 2001, http://web.kitsapsun.com/news/2001/march/03271keyless.html.

5. William R. Graham, Chairman, "Commission to Assess the Threat to the United States From Electromagnetic Pulse (EMP) Attack: Statement before the House Armed Services Committee," July 10, 2008, http://empcommission.org/docs/GRAHAMtestimony10JULY2008.pdf, 3.

6. Ibid., 2.

7. "Pakistani Press Praises Iran's Courage in Confrontation with USA," December 23, 2011, http://www.mathaba.net/news/?x=629629?rss, quoting the *Iran Daily*.

8. Ibid.

9. John A. Brunderman, "High Power Radio Frequency Weapons: A Potential Counter to U.S. Stealth and Cruise Missile Technology," December 1999, http://blockyourid.com/~gbpprorg/mil/herf/brundeman.pdf.

10. Ibid., introduction, 8. (Page numbers listed from this document refer to page numbers in the pdf taskbar, not page numbers as shown on the printed document.)

11. Ibid., vi.

12. Excerpts from the Ranets E product brochure and a technical analysis are posted at the website of Air Power Australia, in a report by Carlo Kopp, titled "Russian/Soviet Point Defence Weapons (Technical Report APA-TR-2008-0502), http://www.ausairpower.net/APA-Rus-PLA-PD-SAM.html.

13. Graham, Statement before the House Armed Services Committee, 4 (see chap. 6, n. 8).

14. See Joby Warrick, "Fear That Nuke Bomb Plan Was Secretly Shared: Report Says Blueprints For Advanced Weapon Were Found on Weapon Smugglers' Computer," *San Francisco Chronicle*, June 15, 2008, http://www.sfgate.com/news/article/Fear-that-nuke-bomb-plan-was-secretly-shared-3209717.php

15. Ibid.

16. Graham, Statement before the House Armed Services Committee, 4.

17. U.S. Army, *Department of Defense Fiscal Year (FY) 2012 Budget Estimates: Justification Book*, vol. 5A, February 2011, http://asafm.army.mil/Documents/OfficeDocuments/Budget/BudgetMaterials/FY12/rforms//vol5a.pdf, 229.

18. Defense Logistics Agency, "Exhibit R-2, RDT&E Budget Item Justification: PB 2011: R-1 Line Item #47," February 2010, http://www.dtic.mil/descriptivesum/Y2011/Other/0603720S_PB_2011.pdf, 4.

19. U.S. Air Force, *Department of Defense Fiscal Year (FY) 2012 Budget Estimates: Justification Book*, vol. 1, February 2011, http://www.saffm.hq.af.mil/shared/media/document/AFD-110211-028.pdf, 35. (Page numbers are as seen on the pdf taskbar.)

20. Ibid., 94.

21. Carafano, Spring, and Weitz, "Before the Lights Go Out," 7. (See chap. 6, n. 11.)
22. "Electromagnetics: Frying Tonight," *Economist*, October 15, 2011, http://www.economist.com/node/21532245.
23. See http://www.apelc.com/system4.html.
24. David Jackson, "Obama: End Afghanistan War, Rebuild USA, *The Oval* (*USA Today*), May 6, 2012, http://content.usatoday.com/communities/theoval/post/2012/05/obama-end-afghanistan-war-rebuild-us/1#.UIeZ6m_A9ZA
25. Carafano, Spring, and Weitz, "Before the Lights Go Out," 10.
26. Wilson, *High Altitude Electromagnetic Pulse (HEMP) and High Power Microwave (HPM) Devices*, 9. (See chap. 3, n. 20.)
27. See http://commdocs.house.gov/committees/security/has204000.000/has204000_0f.htm.
28. Interim Report of the Defense Science Board (DSB) Task Force on the Survivability of Systems and Assets to Electromagnetic Pulse (EMP) and Other Nuclear Weapon Effects (NWE): Summary Report No. 1, Office of the Under Secretary of Defense for Acquisition, Technology, and Logistics, http://www.acq.osd.mil/dsb/reports/ADA550250.pdf, 11.
29. Foster et al., "Report of the Commission: Executive Report," 55. (See chap. 6, n. 6.)
30. William J. Perry, et al., *America's Strategic Posture: The Final Report of the Congressional Commission on the Strategic Posture of the United States* (Washington, D.C.: United States Institute of Peace Press, 2009), http://www.usip.org/files/America's_Strategic_Posture_Auth_Ed.pdf, 111–12. (Page numbers listed are according to the pdf taskbar, not as shown on the printed document.)
31. Wilson, "High Altitude Electromagnetic Pulse (HEMP) and High Power Microwave (HPM) Devices," July 21, 2008, http://www.fas.org/sgp/crs/natsec/RL32544.pdf, 2.
32. Interim Report of the Defense Science Board (DSB) Task Force on the Survivability of Systems and Assets to Electromagnetic Pulse (EMP) and Other Nuclear Weapon Effects (NWE), 5.
33. Ibid., 7.
34. Ibid., 7; Defense Threat Reduction Agency & USSTRATCOM Center for Combating WMD, "About DTRA / SCC-WMD," http://www.dtra.mil/About.aspx.
35. Defense Threat Reduction Agency & USSTRATCOM Center for Combating WMD, "Radiation-Hardened Technology," http://www.dtra.mil/Missions/NuclearDeterrenceDefense/RadiationHardenedTechnology.aspx.
36. Interim Report of the Defense Science Board (DSB) Task Force: Summary Report No. 1, p. 7.
37. Statement by Lieutenant General Robert L. Schweitzer. (See chap. 6, n. 1.)
38. Interim Report of the Defense Science Board (DSB) Task Force, 8.
39. Ibid., 8.
40. Ibid., 8–9.
41. Ibid., 9.
42. Carafano, Spring, and Weitz, "Before the Lights Go Out," 6–7.
43. "Introduction to the Threat," December 13, 2011, at the website After EMP: Surviving Electromagnetic Pulse, http://afteremp.com.

CHAPTER 8: WE CAN STILL DO THE RIGHT THING

1. Carafano, Spring, and Weitz, "Before the Lights Go Out," 10. (See chap. 6, n. 11.)
2. Ibid.
3. Ibid., 11.
4. Ibid.
5. "High-Impact, Low-Frequency Event Risk to the North American Bulk Power System: A Jointly-Commissioned Summary Report of the North American Electric Reliability Corporation and the U.S. Department of Energy's November 2009 Workshop," June 2010, http://www.nerc.com/files/HILF.pdf.

6. See their website at http://www.nerc.com/.

7. "High-Impact, Low-Frequency Event Risk to the North American Bulk Power System," 3, 77.

8. Daniel M. Kammen, "Investing in the Future: R&D Needs to Meet America's Energy and Climate Challenges," Select Committee on Energy Independence and Global Warming, U.S. House, September 10, 2008, http://globalwarming.markey.house.gov/tools/2q08materials/files/0147.pdf, 2.

9. PRNewswire, "SPX Unveils Newly Expanded SPX Transformer Solutions Manufacturing Facility in Waukesha, Wisconsin, April 12, 2012, http://www.prnewswire.com/news-releases/spx-unveils-newly-expanded-spx-transformer-solutions-manufacturing-facility-in-waukesha-wisconsin-147143535.html

10. "Mitsubishi Electric to Build Transformer Factory in Memphis" (news release), Mitsubishi Electric, February 14, 2011, http://www.mitsubishielectric.com/news/2011/0214-a.html.

11. Vergano, "One EMP Burst and the World Goes Dark." (See chap. 3., n. 3.)

12. F. Michael Maloof, "Report Denying EMP Danger Called 'Junk Science,'" *WND*, March 5, 2012, http://www.wnd.com/2012/03/report-denying-danger-emp-danger-called-junk-science/.

13. Ibid.

14. Ibid.

15. Ibid.

16. The Sage Policy Group, "Initial Economic Assessment of Electromagnetic Pulse (EMP) Impact upon the Baltimore-Washington-Richmond Region," 20. (See chap. 3, n. 8.)

17. Ibid., 21.

CHAPTER 9: ONE SECOND AFTER, LITERALLY

1. William R. Forstchen, *One Second After* (New York: Forge, 2009), 20.

INDEX

General Brady tells the real story of America's humanitarian victory in Viet Nam. A story of honor, courage, and faith and the miracles it produces, *Dead Men Flying* is a story no American can afford to ignore.

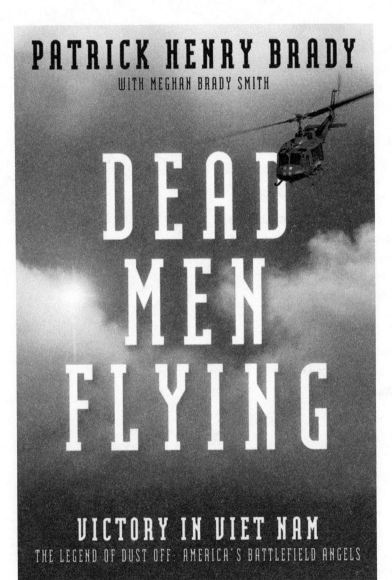

WND Books

WND Books • a WND Company • Washington, DC • www.wndbooks.com

"Stunning ... a wonderful read ... a handbook for li'

Those words of advance praise from another celebrated auth.

scarcely convey just how powerfully mesmerizing is the latest book

by *New York Times* bestselling author and nationally syndicated radio

talk show host Larry Elder.

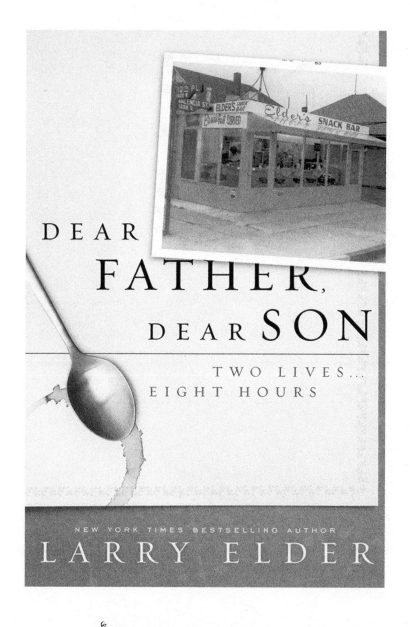

DEAR

FATHER,

DEAR SON

TWO LIVES...
EIGHT HOURS

NEW YORK TIMES BESTSELLING AUTHOR

LARRY ELDER

WND BOOKS

WND Books • a WND Company • Washington, DC • www.wndbooks.com